RESIDUAL CURRENT DEVICES

RESIDUAL CURRENT DEVICES

Selection, Operation, and Testing

STANISLAW CZAPP
Faculty of Electrical and Control Engineering,
Gdańsk University of Technology,
Poland

ELSEVIER
ACADEMIC PRESS
An imprint of Elsevier

Academic Press is an imprint of Elsevier
125 London Wall, London EC2Y 5AS, United Kingdom
525 B Street, Suite 1650, San Diego, CA 92101, United States
50 Hampshire Street, 5th Floor, Cambridge, MA 02139, United States
The Boulevard, Langford Lane, Kidlington, Oxford OX5 1GB, United Kingdom

Library of Congress Cataloging-in-Publication Data
A catalog record for this book is available from the Library of Congress

British Library Cataloguing-in-Publication Data
A catalogue record for this book is available from the British Library

ISBN 978-0-323-89783-9

Publisher: Joseph P. Hayton
Acquisitions Editor: Lisa Reading
Editorial Project Manager: Naomi Robertson
Production Project Manager: Kamesh Ramajogi
Cover Designer: Miles Hitchen

Typeset by STRAIVE, India

Contents

Preface

Safety of persons is the fundamental aspect which has to be considered during the design and construction processes of electrical installations. Protective provisions against electric shock shall take precedence over provisions considering the energy efficiency, functionality, and economic aspects. Nowadays, an inseparable element of the safe low-voltage electrical installations is a residual current device. There is no other protection system and no protective device which has such comprehensive protective properties and such a wide range of applications. The use of residual current devices is not limited to protection in case of a fault (protection against indirect contact) by automatic disconnection of supply. They are the only protective devices that can be effective in the protection of people and animals in the event of direct contact with live parts. They can also be used in protection against fire caused by low-current earth faults and leakage currents to which overcurrent protection devices do not react. Compared with circuit-breakers, and especially compared with fuses, residual current devices have a much more complex structure and, therefore, they are inherently more sensitive to adverse external factors. In consequence, it leads to their increased susceptibility to damage and malfunctions. These factors should be taken into account during residual current devices' selection and utilization.

In this book, the idea for residual current protection has been presented. This idea was born in the early 20th century but applied in practice many years later. The evolution in construction types of residual current devices, which has taken place over decades, is discussed. Types and functional properties of the contemporary residual current devices are described. The principle of operation of their main elements is explained. The main parameters of these devices, from the point of view of their selection and application, are indicated. Special constructions of the protective devices, among other portable and socket-outlets residual devices, are described. The reliability of residual current devices and main sources of their defects are discussed. The rules of backup protection of residual current devices without integrated overcurrent protection are presented. The application of residual current devices in TN, TT, and IT low-voltage systems is commented upon. Requirements of the standards, regarding the obligatory use of residual current devices, are quoted. The principles of application of residual current devices in protection in case of a fault, additional protection, as well as protection against fire, are explained. As extended information related to the use of residual current devices in protection against fire, arc fault detection devices are described, which can be treated as complementary protective devices especially in the case of dangerous series arcing. The issue of residual current devices' operation in circuits with nonsinusoidal currents is commented in detail. Nonsinusoidal earth fault currents are the sources of many problems in residual current devices' operation. Therefore special attention has to be given during their selection to such circuits. The unique results of my own tests of residual current devices' tripping, under distorted earth fault currents, are presented. Residual current devices also make difficulties during an earth fault loop impedance measurement, which is one of the commonly performed measurements to verify the effectiveness of protection against electric shock by automatic disconnection of the supply. The sources of unwanted tripping of the devices, during this type of measurement, are indicated and suitable methods of measurement are shown. Immunity of residual current devices to the impulse leakage currents is discussed. Moreover, coordination of residual current devices with surge protective devices is commented upon. Principles of residual current devices' verifications, examples of their results, as well as fault diagnosis in circuits with these devices are described. At the end of the book, Appendix A presents examples of residual current devices' selection. Additionally, Appendix B covers a list of the standards regarding residual current devices' performance, application, and operation.

This book is useful for students, tutors, and academics, as well as engineers who are involved in the process of design, maintenance, and verification of safety in low-voltage electrical installations.

Terms and definitions

Term	Source	Definition
General terms		
Low voltage	IEV ref 601-01-26	A set of voltage levels used for the distribution of electricity and whose upper limit is generally accepted to be 1000 V for alternating current
Line conductor or **Phase conductor** (deprecated) (identification: **L**)	IEV ref 826-14-09	Conductor which is energized in normal operation and capable of contributing to the transmission or distribution of electric energy but which is not a neutral or mid-point conductor
Neutral conductor (identification: **N**)	IEV ref 826-14-07	Conductor electrically connected to the neutral point and capable of contributing to the distribution of electric energy
Protective conductor (identification: **PE**)	IEV ref 826-13-22	Conductor provided for purposes of safety, for example protection against electric shock Note: In an electrical installation, the conductor identified PE is normally also considered as protective earthing conductor.
Mid-point conductor (identification: **M**)	IEV ref 826-14-08	Conductor electrically connected to the mid-point and capable of contributing to the distribution of electric energy
PEN conductor (identification: **PEN**)	IEV ref 826-13-25	Combining the functions of both a protective earthing conductor and a neutral conductor
PEM conductor (identification: **PEM**)	IEV ref 826-13-26	Conductor combining the functions of both a protective earthing conductor and a mid-point conductor

Continued

Term	Source	Definition
PEL conductor (identification: **PEL**)	IEV ref 826-13-27	Conductor combining the functions of both a protective earthing conductor and a line conductor
Electric shock	IEV ref 195-01-04	Physiological effect resulting from an electric current passing through a human or animal body
Protection against electric shock	IEV ref 826-12-02	Provision of measures reducing the risk of electric shock
Live part	IEV ref 195-02-19	Conductor or conductive part intended to be energized in normal operation, including a neutral conductor, but by convention not a PEN conductor or PEM conductor or PEL conductor
Exposed-conductive-part	IEV ref 195-06-10	Conductive part of equipment which can be touched and which is not normally live, but which can become live when basic insulation fails
Extraneous-conductive-part	IEV ref 195-06-11	Conductive part not forming part of the electrical installation and liable to introduce an electric potential, generally the electric potential of a local earth
Direct contact	IEV ref 195-06-03	Electric contact of persons or animals with live parts
Basic protection	IEV ref 826-12-05	Protection against electric shock under fault-free conditions Note: For low-voltage installations, systems, and equipment, basic protection generally corresponds to protection against direct contact.
Indirect contact	IEV ref 195-06-04	Electric contact of persons or animals with exposed-conductive-parts which have become live under fault conditions
Fault protection	IEV ref 826-12-06	Protection against electric shock under single fault conditions Note: For low-voltage installations, systems, and

Term	Source	Definition
		equipment, fault protection generally corresponds to protection against indirect contact, mainly with regard to failure of basic insulation.
Automatic disconnection of supply	IEV ref 195-04-10	Interruption of one or more of the line conductors effected by the automatic operation of a protective device in case of a fault
Additional protection	IEV ref 826-12-07	Protective measure in addition to basic and/or fault protection Note: Additional protection is generally used in case of special external influences or locations by which under certain circumstances, e.g. careless use of electric energy, a fatal situation maybe avoided or eased.
Alternating current	IEV ref 131-11-24	Electric current that is a periodic function of time with a zero direct component or, by extension, a negligible direct component
Pulsating direct current	IEC 61008-1	Current of pulsating waveform (IEV 101-04-34) which assumes, in each period of the rated power frequency, the value 0 or a value not exceeding 0.006 A DC during one single interval of time, expressed in angular measure, of at least 150°
Smooth direct current	IEC 60755	Direct current which is ripple free
Earth fault current	IEC 61008-1	Current flowing to earth due to an insulation fault
Earth leakage current	IEC 61008-1	Current flowing from the live parts of the installation to earth in the absence of an insulation fault
Current delay angle α	IEC 61008-1	The time, expressed in angular measure, by which the starting instant of current conduction is delayed by phase control

Continued

Term	Source	Definition
Safety extra-low voltage (SELV)	IEV ref 826-12-31 EN 61140	Electric system in which the voltage cannot exceed the value of extra-low voltage: – under normal conditions and – under single fault conditions, including earth faults in other electric circuits
	IEV ref 851-15-08	AC voltage the r.m.s. value of which does not exceed 50 V or ripple-free DC voltage the value of which does not exceed 120 V, between conductors, or between any conductor and reference earth, in an electric circuit which has galvanic separation from the supplying electric power system by such means as a separate winding transformer
Protective extra-low voltage (PELV)	IEV ref 826-12-32 EN 61140	Electric system in which the voltage cannot exceed the value of extra-low voltage: – under normal conditions and – under single fault conditions, except earth faults in other electric circuits
Special terms referred to residual current devices		
Residual current device (RCD)	IEV ref 442-05-02	Mechanical switching device or association of devices designed to make, carry, and break currents under normal service conditions

Term	Source	Definition
In USA: **Ground fault current interrupter (GFCI)**		and to cause the opening of the contacts when the residual current attains a given value under specified conditions
Residual current operated circuit-breaker without integral overcurrent protection (RCCB)	IEV ref 442-05-03	Residual current operated circuit-breaker not designed to perform the functions of protection against overloads and/or short circuits
Residual current operated circuit-breaker with integral overcurrent protection (RCBO)	IEV ref 442-05-04	Residual current operated circuit-breaker designed to perform the functions of protection against overloads and/or short circuits
Residual direct current detecting device (RDC-DD)	IEC 62955	A device to be used for charging of electric vehicles. RDC-DDs are intended to remove or initiate removal of the supply to electric vehicles in cases where a smooth residual direct current equal to or above 6 mA is detected. The value of 6 mA for smooth residual direct current was chosen to prevent impairing the correct operation of an upstream RCD Type A. There are two classes of RDC-DDs: – RDC-MD (monitoring devices), – RDC-PD (protective devices).
Portable residual current device (PRCD)	HD 639 S1	Portable devices for household and similar uses consisting of a plug, a residual current device (RCD), and one or more socket-outlets or a provision for connection, functionally independent of, or functionally dependent on, line voltage;

Continued

Term	Source	Definition
		They are intended to provide protection against shock hazard in case of direct contact, in addition to the protection provided by the fixed installations for the circuit downstream. PRCDs have a rated residual operating current not exceeding 0.03 A.
Socket-outlets residual current protective device (SRCD)	IEC 62640	An RCD incorporated in, or specifically intended for use with two pole socket-outlets, with or without earthing contact for household and similar uses; SRCDs are intended to be used in single-phase systems such as phase to neutral or phase to phase or phase to earthed middle conductor. SRCDs are only intended to provide additional protection downstream of the SRCD. SRCDs are intended for use in circuits where the fault protection (indirect contact protection) is already assured upstream of the SRCD.
Residual current monitor (RCM)	IEC 62020-1	RCMs are intended to monitor the residual current of the installation and to give a warning if the residual current between a live part and an exposed conductive part or earth exceeds a predetermined level.
Circuit-breaker **Miniature circuit-breaker (MCB)**	IEV ref 441-14-20	Mechanical switching device, capable of making, carrying, and breaking currents under normal circuit conditions and also making, carrying for a specified duration, and breaking currents under specified abnormal circuit conditions such as those of short circuit
Arc fault detection device (AFDD)	EN 62606	Device intended to mitigate the effects of arcing faults by

Term	Source	Definition
In USA: **Arc fault circuit interrupter (AFCI)**		disconnecting the circuit when an arc fault is detected
Residual current	IEV ref 826-11-19	Algebraic sum of the values of the electric currents in all live conductors, at the same time at a given point of an electric circuit in an electrical installation
Residual operating current	IEV ref 442-05-20	Value of residual current which causes the residual current device to operate under specified conditions
Rated residual operating current ($I_{\Delta n}$)	IEC 61008-1	The value of residual operating current, assigned to the residual current device by the manufacturer, at which the residual current device shall operate under specified conditions; For a residual current device having multiple settings of residual operating current, the highest setting is used to designate it
Residual nonoperating current	IEV ref 442-05-21	Value of residual current at and below which the residual current device does not operate under specified conditions
Rated residual nonoperating current	IEC 61008-1	The value of residual nonoperating current, assigned to the residual current device by the manufacturer, at which the residual current device does not operate under specified conditions
Making capacity	IEC 61008-1	A value of the AC component of a prospective current that an RCD (RCCB) is capable of making at a stated voltage under prescribed conditions of use and behavior
Conditional residual short-circuit current	IEV ref 442-05-22	Value of the AC component of a residual prospective current which a residual current device without integral short-circuit protection but protected by a short-circuit protective device in series, can withstand under specified conditions of use and behavior

Continued

Term	Source	Definition
Breaking capacity	IEC 61008-1	A value of the AC component of a prospective current that an RCD (RCCB) is capable of breaking at a stated voltage under prescribed conditions of use and behavior
Residual making and breaking capacity	IEV ref 442-05-27	Value of the AC component of a residual prospective current that a residual current device can make, carry for its opening time, and break under specified conditions of use and behavior

Sources

IEV: *International Electrotechnical Vocabulary*

EN 61140: *IEC 61140:2016 Protection against electric shock—Common aspects for installation and equipment*

IEC 61008-1: *IEC 61008-1:2010 Residual current operated circuit-breakers without integral overcurrent protection for household and similar uses (RCCBs)—Part 1: General rules*

IEC 60755: *IEC 60755:2017 General safety requirements for residual current operated protective devices*

IEC 62955: *IEC 62955:2018 Residual direct current detecting device (RDC-DD) to be used for mode 3 charging of electric vehicles*

EN 62606: *IEC 62606:2013 General requirements for arc fault detection devices*

HD 639 S1: *HD 639 S1:2002 Electrical accessories—Portable residual current devices without integral overcurrent protection for household and similar use (PRCDs)*

IEC 62020-1: *IEC 62020-1:2020 Electrical accessories—Residual current monitors (RCMs)—Part 1: RCMs for household and similar uses*

IEC 62640: *IEC 62640:2011 Residual current devices with or without overcurrent protection for socket-outlets for household and similar uses*

CHAPTER 1

Genesis of residual current devices

At the beginning of the 20th century, systems for phase currents summation had already been known and used. In the year 1903 an arrangement of current transformers for detection of earth fault currents was patented by Schuckert, and in the year 1918 Petersen constructed an active power earth fault relay. For generators and power transformers the transverse differential protection, as well as the longitudinal differential protection, had been commonly used.

The first document that refers to a residual current protective system of persons is dated in the year 1928. In this year a patent application appeared in Germany, considering the protection of persons and animals against the effects of direct contact with live conductors in low-voltage power networks [1]. According to the description included in this patent application, tripping of the protective device should be forced by a sum of currents in all live conductors (ger. *Summenstromauslösung*), which is equal to the residual current in the protected circuit (Fig. 1.1).

This description is not precise, it is not explained what mechanism is dedicated to perform the tripping of the protection device. It is only written that the tripping should occur very fast (no more than 0.1 s) and with high sensitivity (ger. *Hochempfindliche Schnellauslösung*), what excludes harmful effect for a person touching a live conductor. It is also emphasized that this protection device reacts to direct touching of a live conductor, in contradiction to the other known means of protection against electric shock.

Considering the idea of the inventor, a protection device should be able to trip out when the residual current of value around 10–50 mA flows. In those days, the performance of a protection device of such parameters (e.g., rated tripping current 30 mA, tripping time up to 0.1 s) was impossible, due to significant leakage currents in installations as well as an insufficient level of electrical apparatus manufacturing technology.

An early construction of a residual current protective device is described in the book "Protective earthing and protection against contact" (1937) authored by Jan Osolsobě (Fig. 1.2) [2]. This protection device,

Residual Current Devices
https://doi.org/10.1016/B978-0-323-89783-9.00011-4

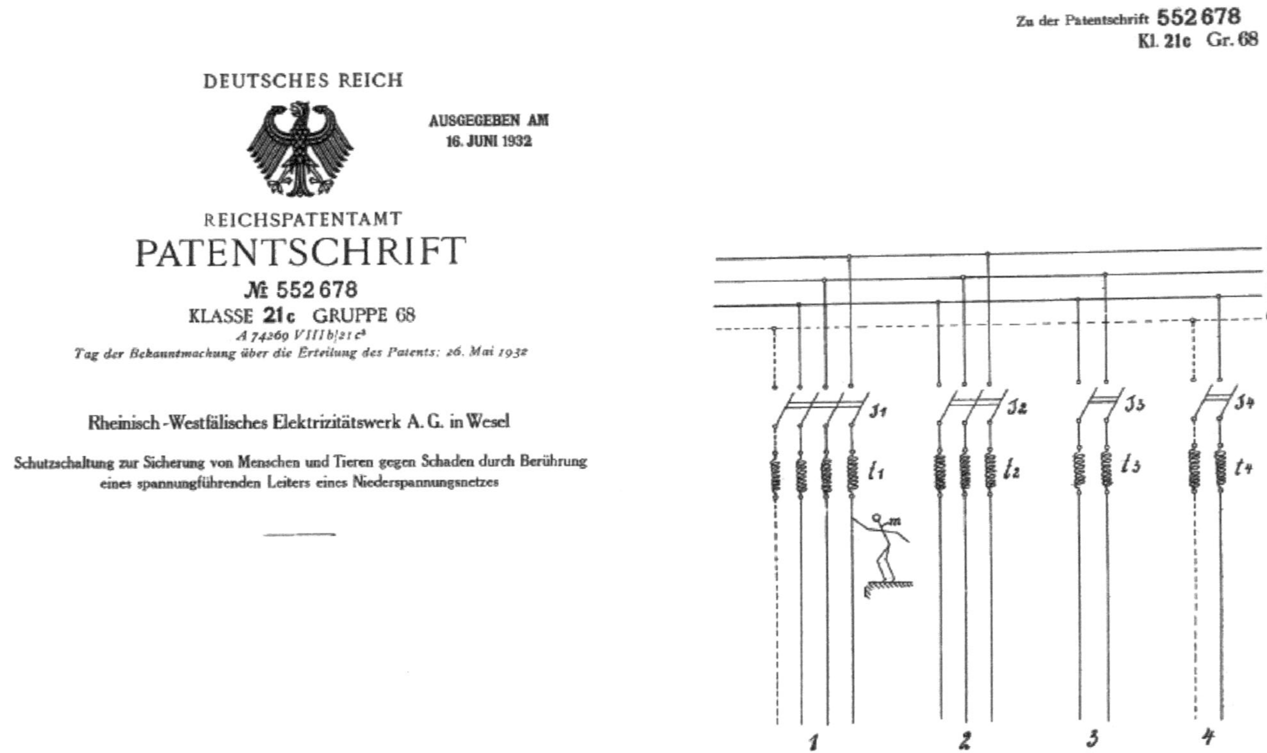
DEUTSCHES REICH

AUSGEGEBEN AM
16. JUNI 1932

REICHSPATENTAMT

PATENTSCHRIFT

№ 552678

KLASSE 21c GRUPPE 68

A 74269 VIII b/21c[3]

Tag der Bekanntmachung über die Erteilung des Patents: 26. Mai 1932

Rheinisch-Westfälisches Elektrizitätswerk A. G. in Wesel

Schutzschaltung zur Sicherung von Menschen und Tieren gegen Schaden durch Berührung eines spannungführenden Leiters eines Niederspannungsnetzes

Zu der Patentschrift 552678
Kl. 21c Gr. 68

Fig. 1.1 Pages 1st and 3rd of the patent application DRP 552678 from 1928, considering the protection of persons and animals in low-voltage networks [1].

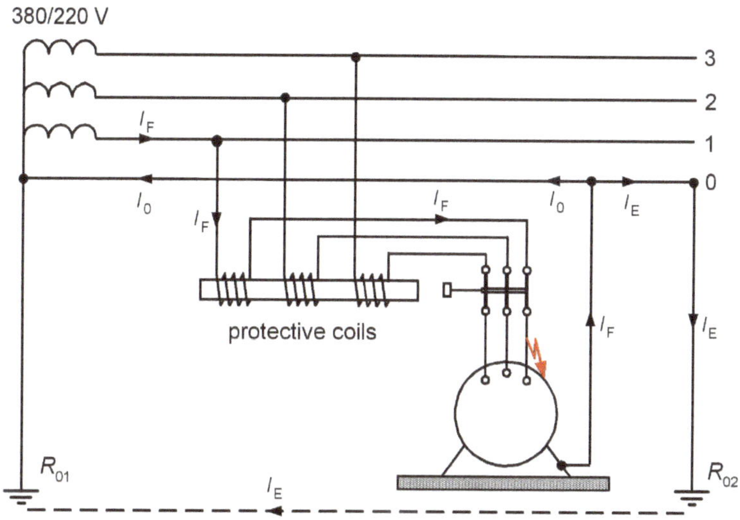

Fig. 1.2 Three-pole residual current device according to J. Osolsobě; I_F—fault current, I_E—earth current, I_0—zero-conductor current, R_{01}—resistance-to-earth of the supplying substation, R_{02}—resistance-to-earth at the end of the network.

supposedly, has rated residual operating current equal to 2 A. As it is presented in Fig. 1.2, there is no current transformer, but there are three protective coils connected in series to the line conductors of the main circuit. A protective conductor is connected to the enclosure of the load. The residual current protective device should operate in case of an earth fault, which enables using a relativity high value of resistance of the earthing system. Therefore the protection device ensured protection in case of a fault (protection against indirect contact) according to the present terminology.

Paul Schnell in his book from the year 1943 [3] described a protective device which looks like residual current devices (RCDs) produced according to the present-day methodology. The described protective device (Fig. 1.3) is composed of a current transformer, an electromechanical relay, a permanent magnet, and a spring. Earth fault current produces secondary current flowing through the electromechanical relay, interacting with the permanent magnet and the spring. This current reduces magnetic flux derived from the permanent magnet, and for a relatively high value of the current, the spring trips the protective device. The rated residual operating current of this device was equal to 10 mA, and tripping time was not higher than 0.1 s.

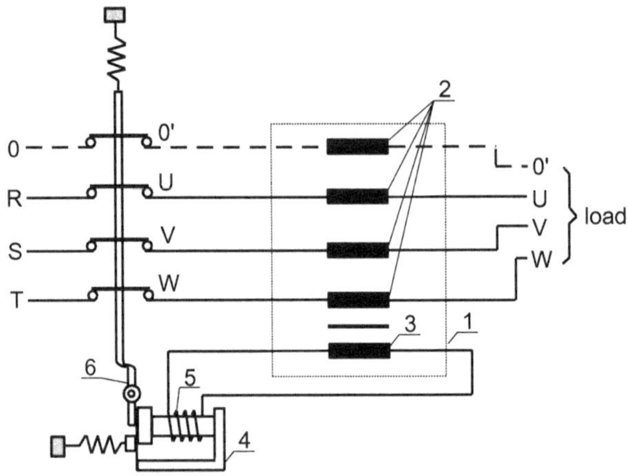

Fig. 1.3 Residual current device according to P. Schnell; 1—current transformer, 2—primary winding of the current transformer, 3—secondary winding of the current transformer, 4—electromechanical relay, 5—coil of the electromechanical relay, 6—triggering/switching mechanism.

The first type of residual current devices, produced on an industrial scale, appeared in the market in 1949 (Schrack), and 1951 (Felten & Guilleaume and Schutzapparategesellschaft Paris&Co). These devices were characterized by relatively big size, low sensitivity (rated residual operating current 1 A or 3 A, exceptionally 0.5 A or 0.3 A), and low reliability, especially insufficient immunity to vibrations. Their cost was rather high. Residual current devices produced at that time were voltage dependent, i.e., detection of residual current and operation of a tripping mechanism were possible only when the auxiliary voltage was delivered to the secondary circuit of the current transformer.

Fig. 1.4 presents an example of the early date construction of a residual current device. After detection of residual current (residual current i_Δ is transformed to the secondary side of current transformer), high sensitivity relay (1) in the secondary circuit of the current transformer (4) affects the overcurrent release (2), which is supplied from the main circuit. For increasing reliability of operation of the tripping mechanism, the overcurrent release (2) is supplied from all line (phase) conductors via a diode bridge.

In the solution presented in Fig. 1.5, a power amplifier (1), supplied from L–N conductors of the main circuit, is used, to amplify the secondary current i_s flowing via the relay (2).

A slightly different solution was proposed in 1957 by prof. Gottfried Biegelmeier, in which a residual current device is equipped with an impulse

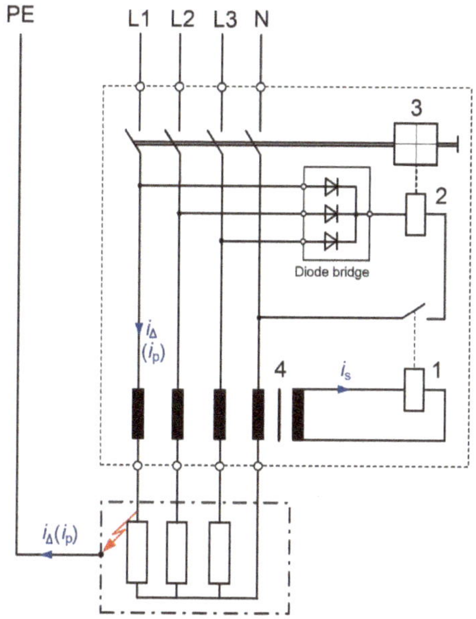

Fig. 1.4 Voltage-dependent residual current device with an overcurrent release in the secondary circuit of the current transformer; i_Δ (i_p)—residual (primary) current of the current transformer, i_s—secondary current of the current transformer. 1—High sensitivity relay, 2—overcurrent release, 3—switching mechanism, 4—current transformer.

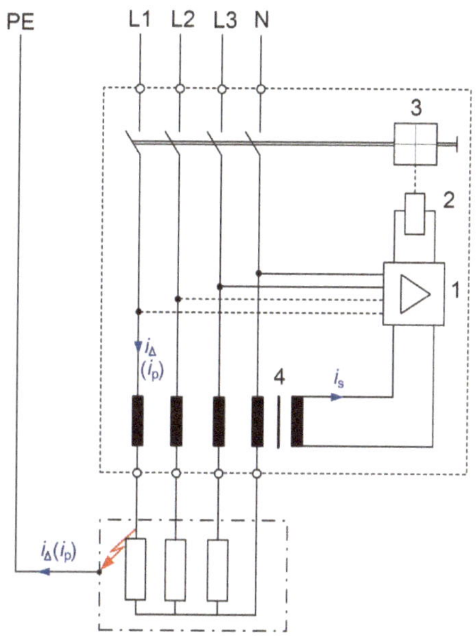

Fig. 1.5 Voltage-dependent residual current device with a power amplifier; i_Δ (i_p)—residual (primary) current of the current transformer, i_s—secondary current of the current transformer. 1—Power amplifier, 2—high sensitivity relay, 3—switching mechanism, 4—current transformer.

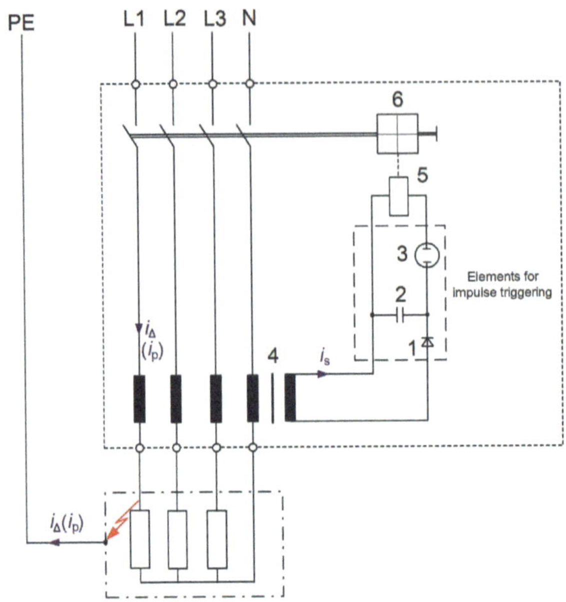

Fig. 1.6 Residual current device with impulse triggering; i_Δ (i_p)—residual (primary) current of the current transformer, i_s—secondary current of the current transformer. 1—Diode, 2—energy storage unit, 3—electronic switch, 4—current transformer, 5—high sensitivity relay, 6—switching mechanism.

triggering/switching mechanism (Fig. 1.6). The secondary current produced by the current transformer (4) energizes an energy storage unit (2), e.g., capacitor. When energy accumulated in this unit is sufficient, an electronic switch (3) (or a lamp in the earlier constructions) turns in the forward-conducting state, and the relay (5) is supplied by the energy storage unit. This construction was a kind of a link between voltage-dependent and voltage-independent residual current devices. Already then, sensitivity was equal to 35 mA. The first residual current device of this type was produced by Felten & Guilleaume in 1958.

A photo of one of the early construction of a residual current device, having rated residual operating current of 30 mA, is presented in Fig. 1.7.

The general principles of construction of the nowadays produced residual current devices are similar to this presented in Fig. 1.6. Most of the residual current devices are voltage independent, composed mainly of an iron core current transformer, an electromechanical relay, and a module

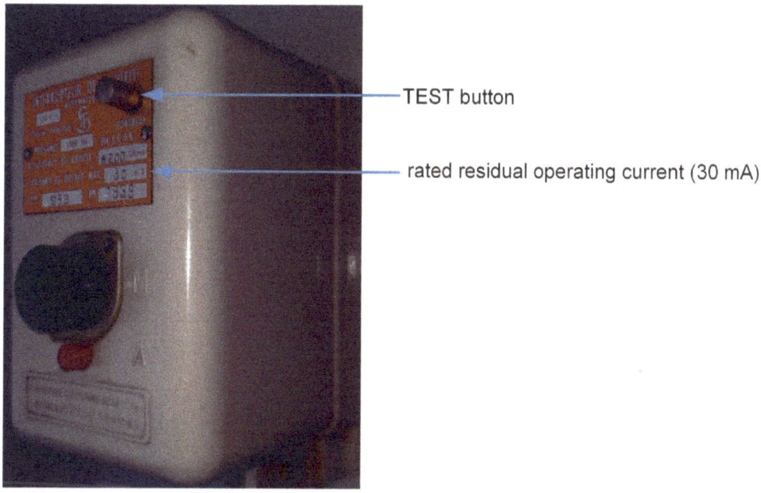

Fig. 1.7 A photo of the early construction of the French residual current device, having rated residual operating current of 30 mA.

with matching electronic elements, which are responsible for intended delay in tripping or improving sensitivity in case of direct pulsating residual current flow (Fig. 1.8). A residual current device should be connected to the live conductors (L1, L2, L3 and neutral N in Fig. 1.8). The protective conductor PE has to go outside the protective device. In case of an earth fault in the protected circuit, the residual current i_Δ is a source of an unbalanced current state in the current sensor (summation current transformer 4) and produces the magnetic flux in its iron core. The magnetic flux produces the secondary voltage, which is a source of the secondary current i_s, flowing through the relay (2). If the secondary current i_s reaches a predetermined value, the tripping of the residual current device occurs.

The equivalent circuit of the simplest construction of the RCD is presented in Fig. 1.9A. Such a residual current device is provided for detection of AC residual currents. An electromechanical relay (ER) is connected to the secondary winding of the RCD, without using a module comprising matching electronic elements. In order to increase the sensitivity of RCDs to AC currents, a series capacitor is installed in the secondary circuit (Fig. 1.9B).

This capacitor is also useful for detecting DC pulsating currents. Even better for DC pulsating currents detection is a solution presented in

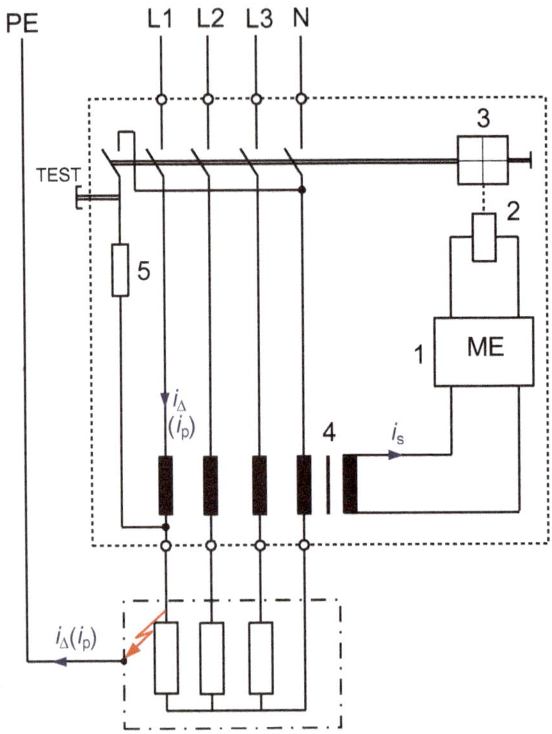

Fig. 1.8 Voltage-independent, modern residual current device; i_Δ (i_p)—residual (primary) current of the current transformer, i_s—secondary current of the current transformer. 1—Matching electronics elements (ME), 2—relay, 3—switching mechanism, 4—current transformer, 5—resistor for current limiting in a testing circuit, TEST—test button.

Fig. 1.9C. In this solution, two resonance capacitors, serial and parallel, are installed in the secondary circuit. In order to achieve selective tripping of two RCDs installed in series, the upstream RCD must be of the selective (delayed) type. Delayed tripping of RCDs is achieved by the application of matching elements, as presented in Fig. 1.9D.

Fig. 1.10 presents a view of the nowadays produced residual current device ($I_{\Delta n}$ = 30 mA). It is a voltage-independent protective device which detects residual sinusoidal AC waveforms and pulsating DC waveforms (A-type residual current device).

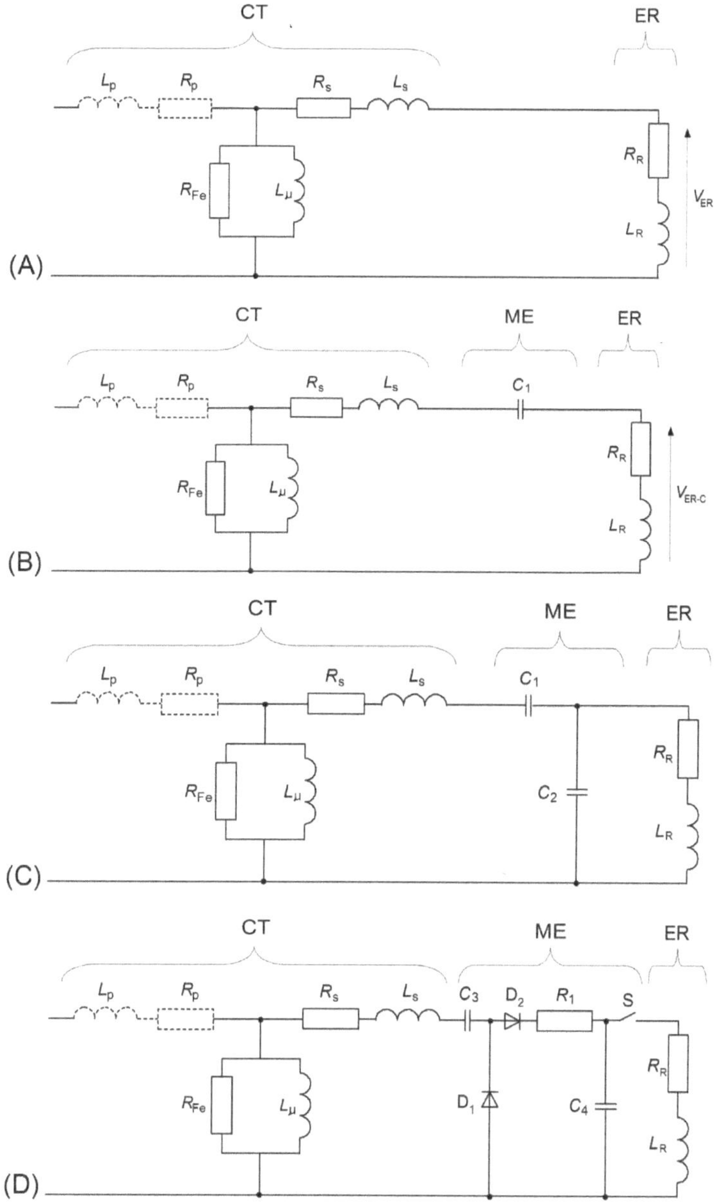

Fig. 1.9 Equivalent circuit of residual currents devices: (A) for AC currents detection, (B, C) for AC currents detection and DC pulsating currents detection, (D) with delay in tripping; L_p—inductance of the primary winding, R_p—resistance of the primary winding, L_s—inductance of the secondary winding, R_s—resistance of the secondary winding, L_μ, R_{Fe}—equivalent parameters of the iron core, C_1—series resonance capacitor, C_2—parallel resonance capacitor, C_3—capacitor for voltage doubling, C_4—filter capacitor, R_1—load resistance, D_1, D_2—rectifier, S—threshold value switch. CT—current transformer, ME—matching electronic elements, ER—electromechanical relay, V_{ER}—electromechanical relay voltage, $V_{ER\text{-}C}$—electromechanical relay voltage with a series capacitor.

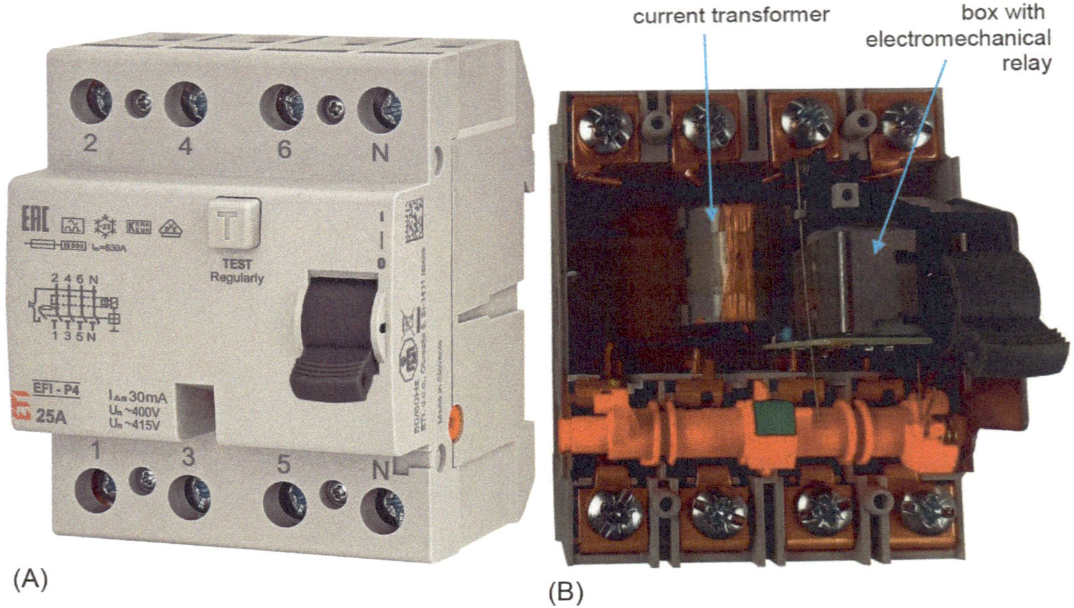

Fig. 1.10 Currently produced residual current device having rated residual operating current 30 mA: (A) general view, (B) internal view,

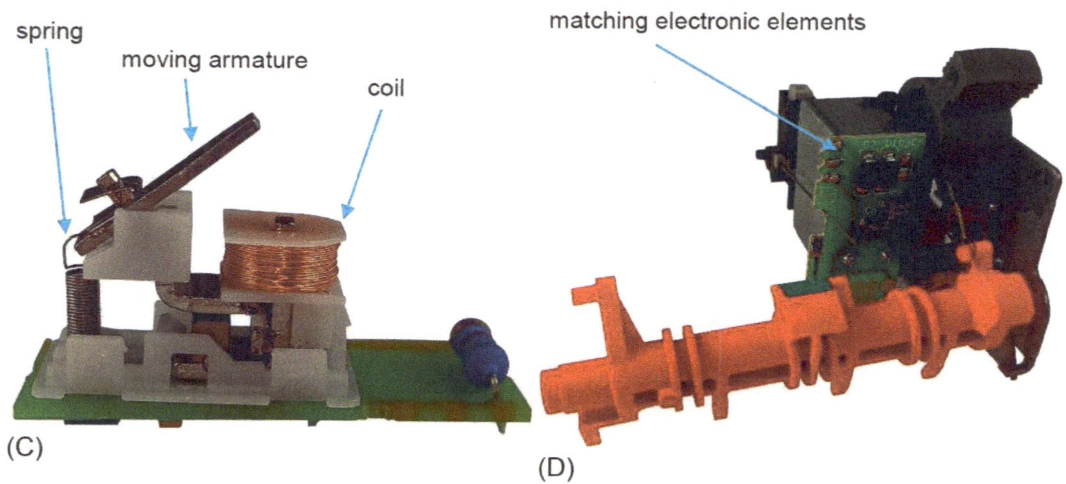

Fig. 1.10, cont'd (C) electromechanical relay, (D) module with matching electronic elements (manufacturer ETI).

References

[1] Deutsches Reichspatent 552678, Schutzschaltung zur Sicherung von Menschen und Tieren gegen Schäden durch Berührung eines spannungsführenden Leiters eines Niederspannungsnetzes, Patentinhaber, Rheinisch-Westfälische Elektrizitäts-werke AG in Wesel, April 1928.
[2] J. Osolsobě, Zemnění a ochrana před dotekem, Elektrotechnický Svaz Československý, Praha, Leden, 1937.
[3] P. Schnell, Differentialschutzschaltung und Lebens- und Brandgefahren in Drehund Wechselstromanlagen, ETZ 9/10 (1943) 119–123.

CHAPTER 2

Construction of modern residual current devices

2.1 Electromechanical relay

One of the main challenges faced by inventors of residual current devices was the development of the structure of the relay, installed in the secondary circuit of the current transformer, which could operate under extremely low power delivered to it. This is the condition of tripping of RCDs without auxiliary voltage. One must be aware that residual current equal to 30 mA delivers a level of power approx. to a fraction of milliwatt to the secondary circuit of RCDs. This is extremely low power. Therefore relays utilized to initiate the tripping of RCDs are closed in normal operation conditions, and they are kept in the closed position by a permanent magnet. Residual current, supported by a spring, generates a force higher than the keeping force of the permanent magnet, and it changes the position of the moving element of the relay. The opposite operation—pulling the moving element from the open position, like in an electromechanical contactor—is impossible, due to very low power produced by the residual current.

A residual current device may contain one of two types of relays. Both are of the electromechanical type, but the first is a **polarized relay** (Fig. 2.1) and the second is a **nonpolarized relay** (Fig. 2.3).

After closing manually an RCD with a polarized relay, when no residual current flows in the main circuit, the moving armature (4) is pulled to the yoke (1). A permanent magnet producing magnetic flux $\phi_{\text{N-S}}$ overcomes the force derived from the spring (5) and keeps this armature in a closed position. When a residual current occurs, the current transformer produces secondary current I_s flowing through the relay, and this current is a source of magnetic flux ϕ_{Is}. At this time, magnetic flux ϕ_{Is} in a half period of the sine wave increases, and in the next half period of the sine wave reduces constant magnetic flux $\phi_{\text{N-S}}$ produced by the permanent magnet. If resultant force coming from both these magnetic fluxes ($\phi_{\text{N-S}} - \phi_{Is}$) is lower than the force of

Residual Current Devices
https://doi.org/10.1016/B978-0-323-89783-9.00005-9

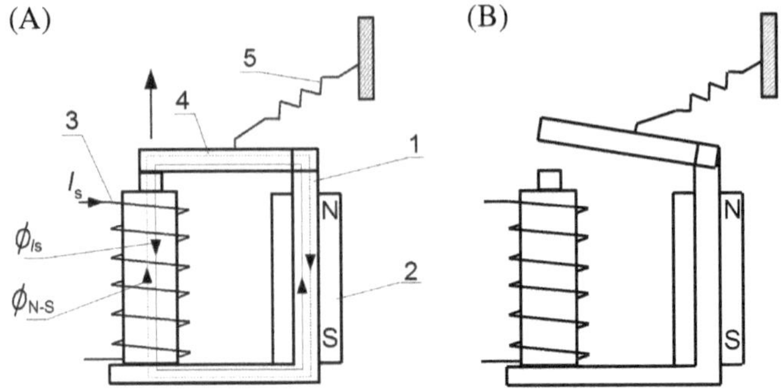

Fig. 2.1 Polarized electromechanical relay of a residual current device: (A) before tripping, (B) after tripping; 1—yoke, 2—permanent magnet, 3—relay winding, 4—moving armature, 5—spring, ϕ_{Is}—magnetic flux from the residual current (current transformed to the secondary side of the current transformer), $\phi_{N\text{-}S}$—magnetic flux from the permanent magnet, I_s—secondary current.

the spring, the moving armature changes its position, and polarized electromechanical relay trips the RCD.

Magnetic force $F_{N\text{-}S}$ derived from magnetic flux $\phi_{N\text{-}S}$ of the permanent magnet is expressed by the following dependency:

$$F_{N-S} = \frac{\phi_{N-S}^2}{s \cdot \mu_0} \tag{2.1}$$

where

$\phi_{N\text{-}S}$ is magnetic flux produced by a permanent magnet,

s is the cross-sectional area of a pole piece,

μ_0 is magnetic constant (vacuum permeability),

whereas resultant force affecting the relay in a state without residual current can be described as follows:

$$F_{N-S} - F_{sp} = \frac{\phi_{N-S}^2}{s \cdot \mu_0} - F_{sp} \tag{2.2}$$

where

F_{sp} is separating force of the spring.

Sinusoidal secondary current i_s (I_s rms), which comes from a residual current i_Δ (I_Δ rms), generates sinusoidal magnetic flux ϕ_{Is} and tries to reduce magnetic flux $\phi_{N\text{-}S}$. In consequence, a resultant magnetic force F_{res} in a reducing half period of a sine wave of the secondary current is described by the dependence:

$$F_{res} = \frac{\phi^2{}_{N-S} - \phi^2{}_{Is}}{s \cdot \mu_0} - F_{sp} \tag{2.3}$$

The simplified visualization of the interaction between the above-mentioned elements is presented in Fig. 2.2. When the waveform F_{res} crosses the line of constant value $F_{N\text{-}S} - F_{sp}$, tripping of the polarized relay occurs (point P_{tr} in Fig. 2.2).

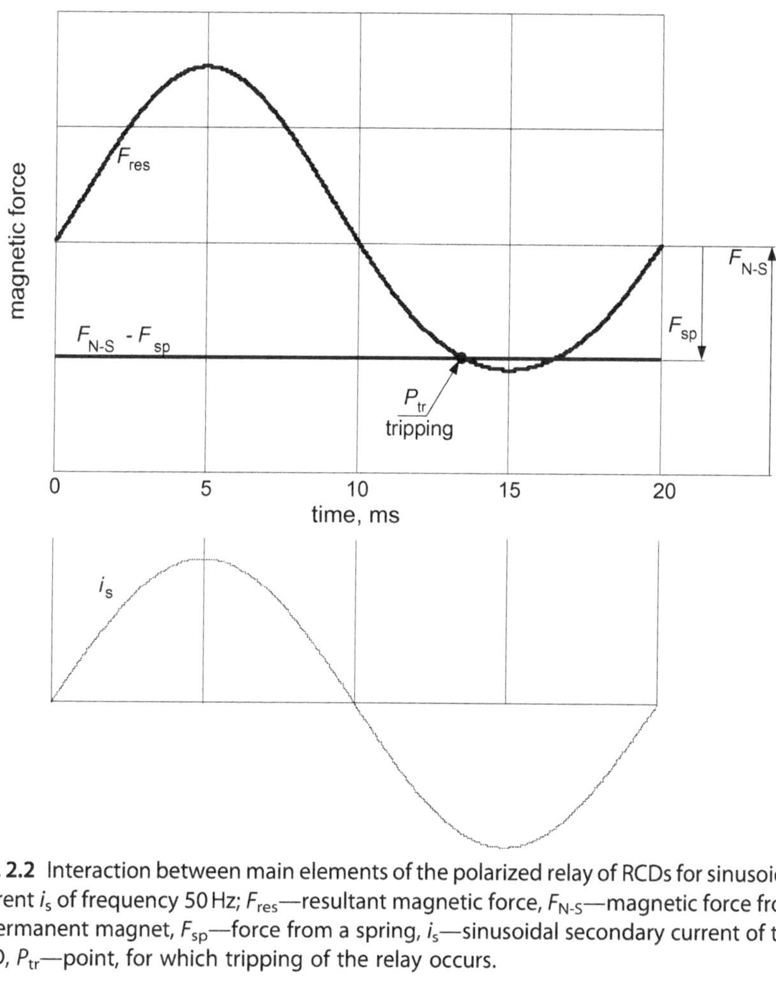

Fig. 2.2 Interaction between main elements of the polarized relay of RCDs for sinusoidal current i_s of frequency 50 Hz; F_{res}—resultant magnetic force, $F_{N\text{-}S}$—magnetic force from a permanent magnet, F_{sp}—force from a spring, i_s—sinusoidal secondary current of the RCD, P_{tr}—point, for which tripping of the relay occurs.

The name of the relay is the "polarized relay" because the direction of the secondary current (positive half-wave vs. negative half-wave), which interacts with the permanent magnet, influences the operation of the relay. Direction (polarization) of the secondary current also influences the time of its operation.

A nonpolarized relay (Fig. 2.3) of RCDs, similarly to the polarized relay, contains yoke (1), permanent magnet (2), relay winding (3), moving armature (4), and spring (5). However, another phenomenon is utilized to make the operation of this relay. When an RCD is closed and no residual current appears, a magnetic flux $\phi_{N\text{-}S}$ produced by the permanent magnet flows through the yoke and the moving armature. After residual current occurs, the secondary current I_s starts to flow in the relay winding. This current produces magnetic flux ϕ_{Is}, whose aim is to block the flux path (in the yoke narrowing) of the magnetic flux $\phi_{N\text{-}S}$. In narrow areas of the yoke, magnetic permeance and permeability decrease. Therefore the mentioned path is blocked. In consequence, the magnetic flux $\phi_{N\text{-}S}$ does not reach the moving armature, and the spring pulls it effectively, causing tripping of the relay.

The name of the relay is the "nonpolarized relay" because direction (polarization) of the secondary current and direction of the magnetic flux in the yoke do not influence the tripping of this relay.

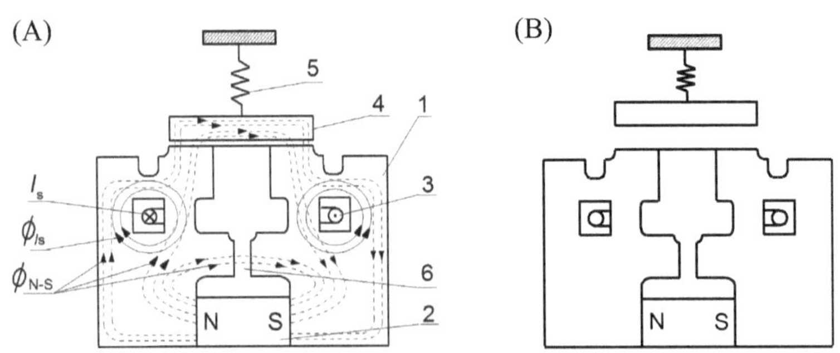

Fig. 2.3 Nonpolarized electromechanical relay of a residual current device: (A) before tripping, (B) after tripping; 1—yoke, 2—permanent magnet, 3—relay winding, 4—moving armature, 5—spring, 6—magnetic shunt, ϕ_{Is}—magnetic flux from the residual current (transformed to the secondary side of the current transformer), $\phi_{N\text{-}S}$—magnetic flux from the permanent magnet, I_s—secondary current.

2.2 Current transformer

Residual current devices contain an iron core summation current transformer, whose aim is to detect residual current, coming from an earth fault or, what is extremely dangerous, from direct contact of a person with live parts. The most often used residual current devices have a rated residual operating current equal to 30 mA. On the other hand, their rated current is relatively high, e.g., 40 or 63 A. The number of primary windings is reduced to a single winding—this is necessary to avoid unwanted power losses, a significant voltage drop across the device, and to limit current transformer dimensions. The working point of the iron core should be located close to the initial permeability range [1]. Manufactures of materials for iron cores of RCD current transformers refer in their catalogs to magnetic field strength 4 mA/cm (parameter μ_4) and/or 15 mA/cm (parameter μ_{15}). Moreover, the cores should be characterized by a very low magnetizing current to achieve a magnetization field strength for the rated residual operating current as high as possible.

Iron cores of RCD current transformers are made of soft magnetics, amorphic Fe-Ni alloys, and nanocrystalline alloys. An example composition of materials is presented in Table 2.1.

Tables 2.2 and 2.3 present materials used for current transformers and present their properties. Ultraperm is a crystalline material which achieves very high permeability. As has been mentioned, the most important is permeability for very low values of magnetic field strength. Fig. 2.4 presents the variation of the permeability for Ultraperm 250 within the wide range of magnetic field strength. Maximal permeability is around 450,000—this value already falls on 10 mA/cm. Moreover, for 4 mA/cm, the permeability exceeds 200,000. It ensures production of a sufficient signal in the secondary winding of the current transformer, even in case of a small value of residual current.

If a residual current device is required to be able to detect DC pulsating currents (A-type RCD), the key element is in the shape of a hysteresis loop

Table 2.1 Composition of selected materials used for cores of RCD current transformers; according to [2].

Current transformers for RCDs	Material components
AC-type, high sensitivity	Magnifer 7904: 80% Ni, 5% Mo, 15% Fe
AC-type, low sensitivity	Magnifer 53: 55% Ni, 45% Fe
	Magnifer 50: 48% Ni, 52% Fe

Table 2.2 Materials used for cores of current transformers of AC-type RCDs; according to [1].

Rated residual operating current of RCDs $I_{\Delta n}$, mA	Material	Density g/cm^3	Saturation induction T	Permeability μ_{15}
10–100	Ultraperm 200/250	8.7	0.74	200,000–300,000
	Ultraperm 10	8.7	0.74	100,000–200,000
10–100 with series capacitor	Ultraperm F60	8.7	0.74	125,000
	Vitroperm 800F	7.35	1.2	145,000
300–500	Permax M	8.25	1.5	50,000
	Permenorm 5000 H2	8.25	1.55	12,000

Table 2.3 Materials used for cores of current transformers of A-type RCDs; according to [1].

Rated residual operating current of RCDs $I_{\Delta n}$, mA	Material	Density (g/cm^3)	Saturation induction T	Permeability μ_{15}
10–100	Vitroperm 800F	7.35	1.2	135,000–160,000
	Ultraperm F80	8.70	0.74	110,000
300–500	Vitroperm 500F	7.35	1.2	20,000–80,000

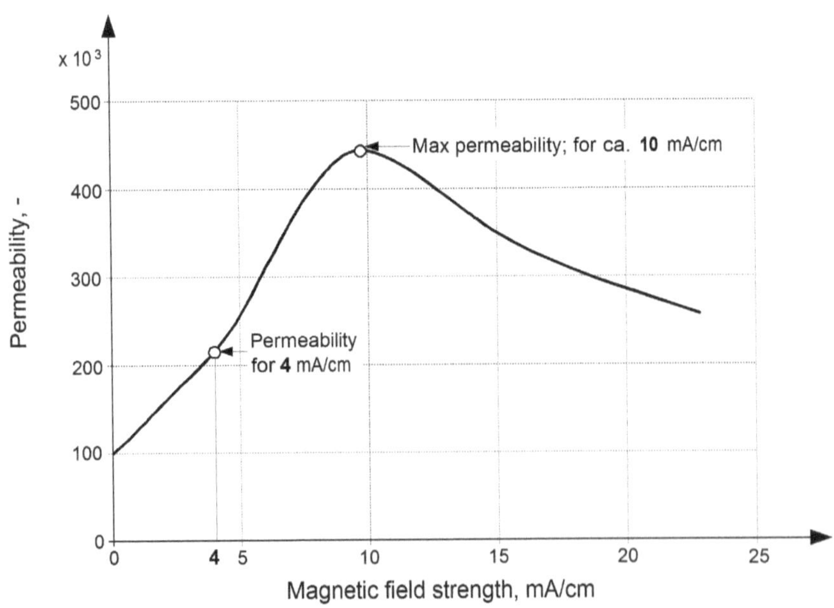

Fig. 2.4 Relative permeability of Fe-Ni alloy Ultraperm 250 (manufacturer Vacuumschmelze) as a function of the magnetic field strength.

of the core, which depends on the chemical components of the core and its production technology. Vitroperm (Table 2.3) is a soft magnetic material which enables detection of DC pulsating currents. Such material gives sufficient variation of the magnetic induction in case of unidirectional pulsating residual currents.

Magnetic cores for residual current devices can be classified into three groups with reference to the ratio of B_{rem}/B_{sat} (remanence induction/saturation induction), as presented in Table 2.4.

The simplest cores, dedicated for the detection of AC residual currents only (AC-type RCD), are characterized by rectangular or round hysteresis loop (Fig. 2.5A and B). When this type of core is applied in an RCD, it is impossible to detect direct current even with relatively high pulsation. As far as AC residual current is concerned, the induction changes are large. When a pulsating DC current flows, a very small change in induction $\Delta B \approx B_{sat} - B_{rem}$ does not guarantee sufficient induced secondary voltage in the secondary winding of the current transformer. To detect such a type of current, a magnetic core characterized by flat hysteresis loop has to be used (Fig. 2.5C).

Table 2.4 Classification of magnetic cores with reference to the shape of magnetic hysteresis loops.

Name of hysteresis loop	B_{rem}/B_{sat}
Rectangular	≈ 0.9
Round	0.6–0.7
Flat	<0.5

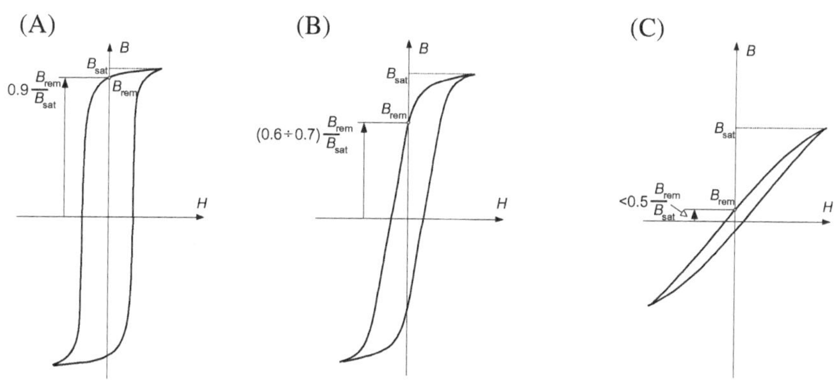

Fig. 2.5 Hysteresis loop: (A) rectangular, (B) round, (C) flat; B_{sat}—saturation induction, B_{rem}—remanence induction; H—magnetic field strength.

For such a type of hysteresis loop, remanence induction B_{rem} is relatively low, and DC current with high pulsation produces high enough change in induction $\Delta B \approx B_{sat} - B_{rem}$.

Fig. 2.6 presents a comparison of the induced secondary voltage for two types of current transformers:

- Current transformer for AC currents (rectangular hysteresis loop)—Fig. 2.6B,
- Current transformer for AC currents and pulsating DC currents (flat hysteresis loop)—Fig. 2.6C.

When pulsating DC current flows, no induced secondary voltage occurs in the secondary winding of the current transformer characterized by rectangular hysteresis loop (Fig. 2.6B). Current transformer characterized by flat hysteresis loop (Fig. 2.6C) gives a relatively high value of this voltage.

It is worth mentioning that in the case of DC pulsating residual currents, the configuration of the secondary circuit of the current transformer is also very important. A series capacitor in this circuit, as presented in Fig. 1.9B, is applied for improving parameters of the secondary voltage of the current transformer. This capacitor cumulates an electric charge, which flows back during the dead time of the DC pulsating residual current waveform. Thanks to this phenomenon, relatively high variation of magnetic induction is ensured, the iron core of the current transformer is better demagnetized, what positively influences waveform of the voltage across the electromechanical relay of the RCD (Fig. 2.7).

Nowadays produced RCDs intended for the detection of AC sinusoidal currents and DC pulsating currents are usually voltage independent. It means that no auxiliary voltage is necessary to ensure operation of the RCD. It is different in case of the necessity of smooth DC current detection. No pulsation or slight pulsation of residual current enforces the introduction of the auxiliary voltage to the secondary circuit of the current transformer. Moreover, some manufacturers utilize two current transformers in RCDs: the first is responsible for the detection of AC sinusoidal current and DC pulsating current; the second is responsible for the detection of smooth DC current and DC current with slight pulsation.

Fig. 2.8 presents a two-core RCD intended for the detection of AC, pulsating DC, and smooth DC residual currents [4,5]. The current transformer marked as 4A is responsible for the detection of AC currents and pulsating DC currents, and no auxiliary voltage is needed for its proper operation. In turn, the second current transformer (4B) and the relay (1) detect smooth DC current (DC current with slight pulsation as well) and requires auxiliary

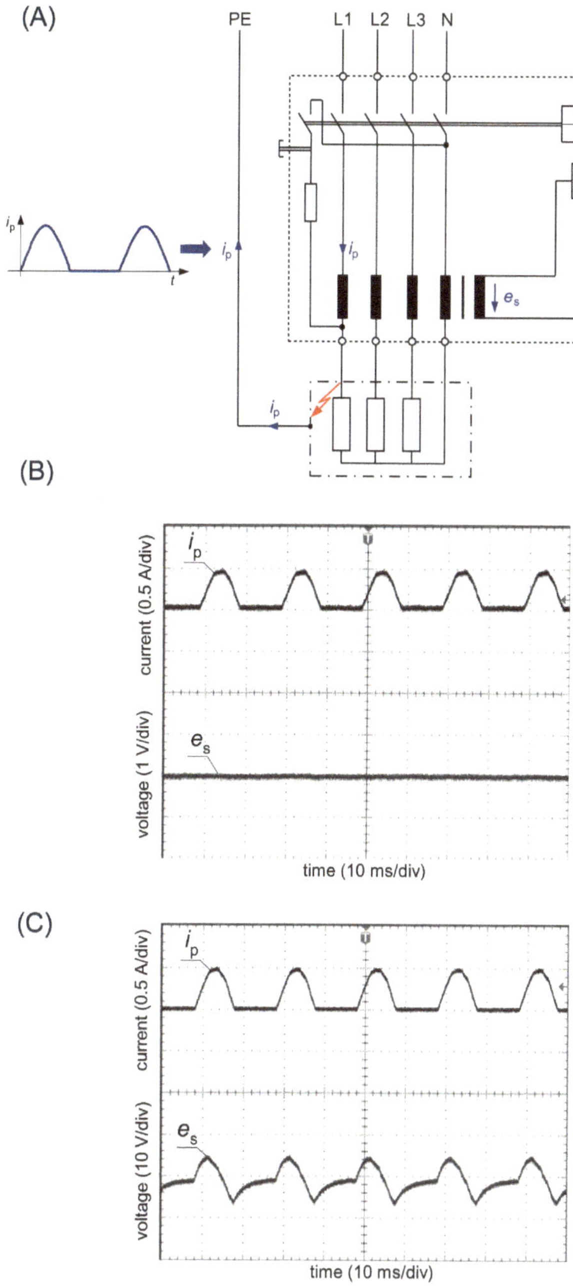

Fig. 2.6 Induced secondary voltage $e_s(t)$ in the secondary winding of the current transformer for half-wave primary current $i_p(t)$: (A) structure of the RCD circuit, (B) response of the current transformer designed for AC currents (rectangular hysteresis loop), (C) response of the current transformer designed for AC currents and pulsating DC currents (flat hysteresis loop).

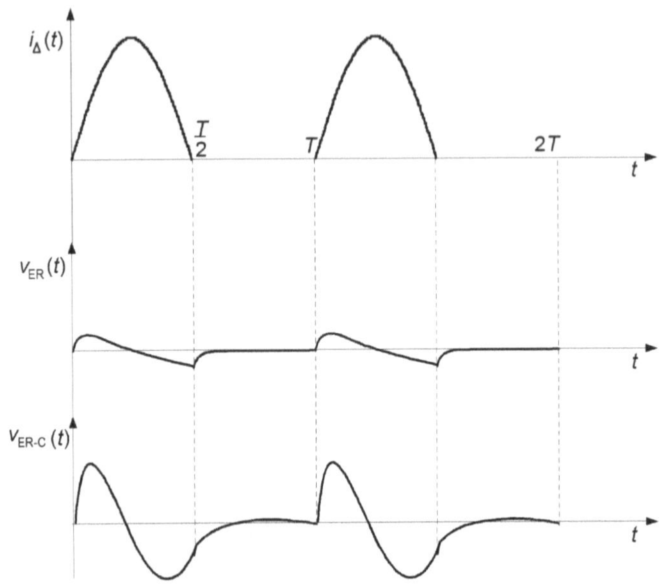

Fig. 2.7 Voltage waveform across the electromechanical relay ER (see Fig. 1.9A and B), in case of pulsating half-wave residual (primary) current $i_\Delta(t)$: $v_{ER}(t)$—voltage without series capacitor, $v_{ER\text{-}C}(t)$—voltage with series capacitor; according to [3].

voltage. In order to increase the reliability of the operation of the system, the auxiliary voltage is delivered to the relay (1) from all line conductors.

The system for detection of smooth DC current operates as follows (Fig. 2.8B). Generator *Gen* produces high-frequency testing current I_{Gen}. This current flows through resistance R which gives voltage drop ΔU_{Gen} across it. If there is no smooth DC residual current $I_{\Delta\text{-DC}}$, the voltage ΔU_{Gen} does not exceed a predetermined level and RCD is still closed. When the $I_{\Delta\text{-DC}}$ flows, the inductance L_{4B} decreases, due to lower permeability of the secondary winding of the transformer (DC current magnetizes the iron core of the current transformer 4B). As a result, both the testing current I_{Gen} and the voltage drop ΔU_{Gen} rise. If the voltage drop exceeds a predetermined level, tripping of the RCD occurs.

2.3 Internal testing circuit

Residual current devices which stay for a relatively long time (many months, years) in the closed position may have a problem with the operation when a residual current of value near to the rated residual operating current ($I_{\Delta n}$)

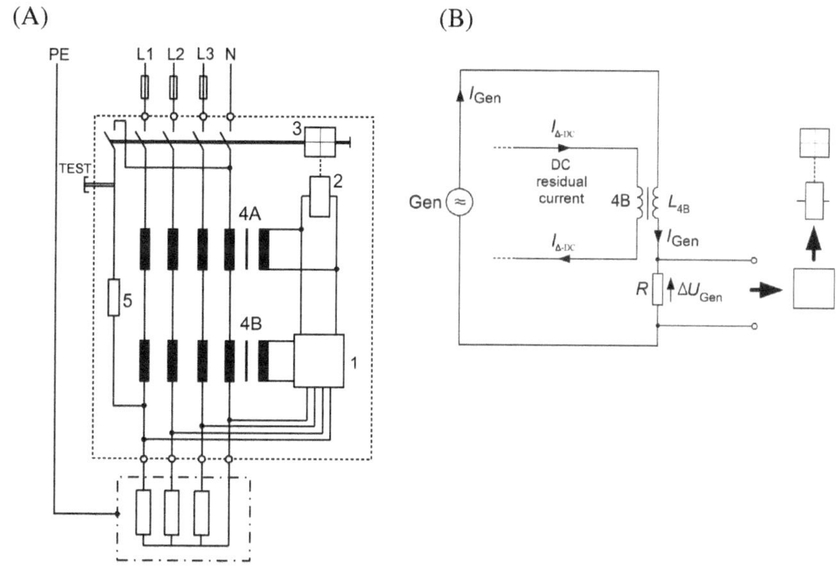

Fig. 2.8 Two-core residual current device for detection of AC, pulsating DC, and smooth DC residual currents: (A) general diagram; (B) example system for detection of smooth DC current. 1—Electronic relay for detection of smooth DC current, 2—tripping relay, 3—switching mechanism, 4A—current transformer for detection of AC and pulsating DC currents, 4B—current transformer for detection of smooth DC current, 5—resistor for current limiting in a testing circuit, TEST–test button, $I_{\Delta\text{-DC}}$—smooth DC current, L_{4B}—inductance of the secondary winding of the transformer 4B, R—reference resistor, ΔU_{Gen}—voltage drop, I_{Gen}—high-frequency testing current, Gen—high-frequency generator.

occurs. However, after opening and closing the RCD manually, as well as after the second, third, or next tripping, its sensitivity comes back to the proper tripping level. This phenomenon is caused by adhesive sticking of the moving armature to the yoke. In order to avoid such a negative phenomenon, every RCD should be periodically tested with the use of its internal testing circuit. Such testing is performed by pushing the TEST button. It was initially recommended to perform this test monthly, but nowadays rather every 6 months. Fig. 2.9 presents the former type of the testing circuit (according to [6]) and Fig. 2.10 presents various configurations of this circuit in contemporary produced RCDs.

Construction of the testing circuit of early RCDs (Fig. 2.9) significantly differs from contemporary solutions. The testing circuit had its own winding

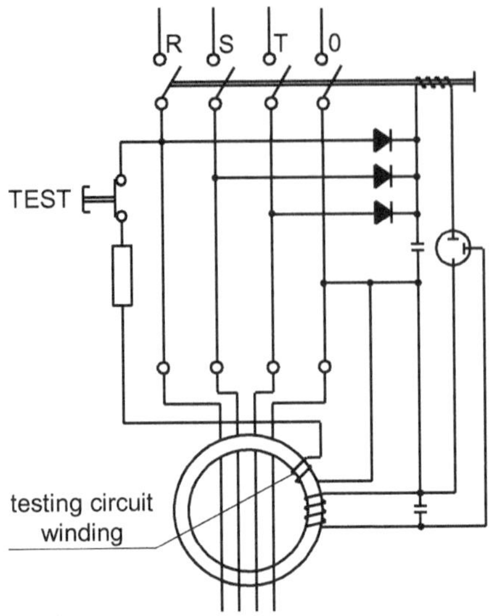

Fig. 2.9 Early type of RCD's internal testing circuit; according to [6].

on the iron core of the current transformer. It leads to increasing the core dimensions as well as to unfavorable power loss production (heat generation).

In order to force testing current inside the RCD, no additional winding on the iron core is necessary. The rule is to connect terminals of the testing circuit—in contrary to the early solution—at opposite sides of the current transformer (Fig. 2.10). Besides this, the testing circuit contains an auxiliary make contact, which opens itself when the RCD opens. This is a protection against the risk of overheating (burning) of the resistor R_T in certain conditions (frequently pushing of the TEST button in short intervals).

A residual current device should be installed in such a way as to ensure the operation of the internal testing circuit. In case of a 2-pole RCD for a single-phase circuit, there is no risk of improper connection. It is different in the case of a 4-pole RCD. When a 4-pole RCD is utilized in a single-phase circuit, a supplying line has to be connected to the poles with the testing circuit. In the case of the RCD from Fig. 2.10A a supplying line has to be connected to poles L1–N, but in the case of the RCD from Fig. 2.10B—to poles L1–L3. A problem with operation of a 4-pole RCD may also occur in a three-wire circuit for supplying a three-phase symmetrical load (Fig. 2.10C). Connection of this circuit to the poles L1–L2–L3 results in no operation of the testing circuit. The remedy for this problem is to connect an additional resistance R_{add} to the poles L2–N, as presented in Fig. 2.10D.

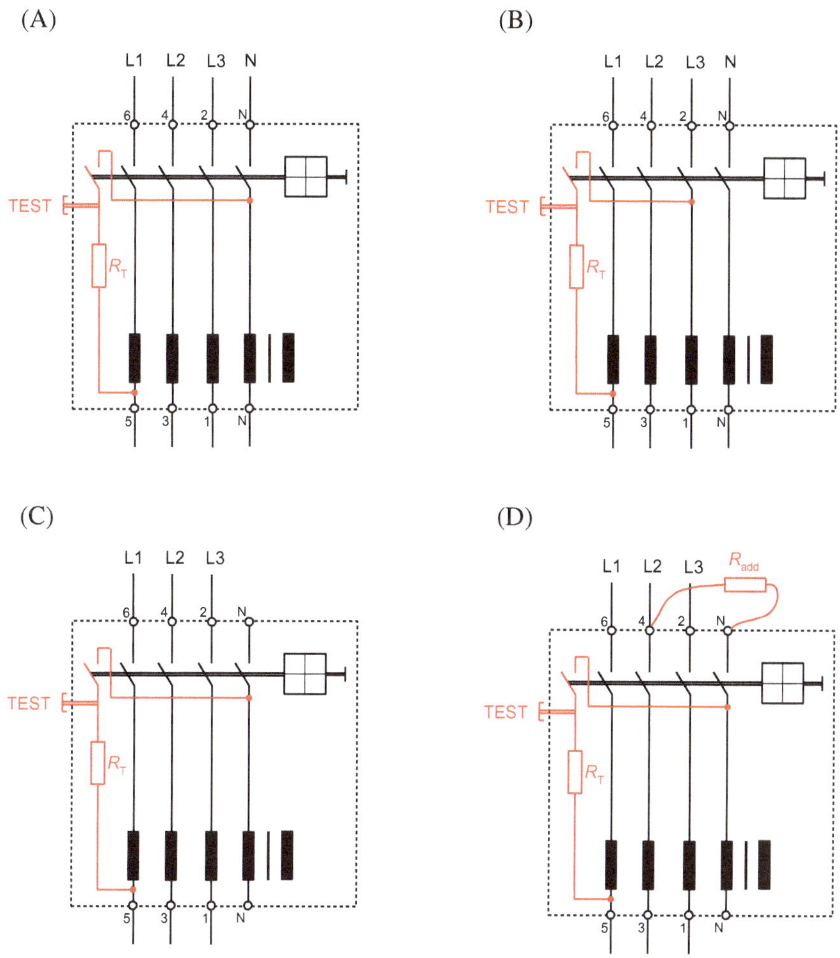

Fig. 2.10 Connections of the internal testing circuit inside the RCD and its coordination with various types of power supply. TEST–test button, R_T—resistance limiting the testing current, R_{add}—additional external resistance for current limiting, connected by qualified personnel. Testing circuit between L1–N and four-wire power supply (A), testing circuit between L1–L3 and four-wire power supply (B), testing circuit between L1–N and three-wire power supply—no proper operation of the testing circuit (C), testing circuit between L1–N and three-wire power supply—the modification to ensure proper operation of the testing circuit (D).

During testing operation of the RCD by pushing the TEST button, the current value in the testing circuit is around $2.5I_{\Delta n}$. This is only a rough operational test and, therefore, it is not possible to state whether the RCD is fully functional and whether the RCD operates in case of the occurrence of the residual current equal to $I_{\Delta n}$. The internal testing circuit should operate properly for supplying voltage within the range (0.85–1.10) V_{ns}, where V_{ns} is a nominal voltage of the testing circuit.

References

[1] Vacuumschmelze, Tape wound cores for (RCCB) residual current circuit breaker, Vacuumschmelze Data (2003).

[2] VDM, Weichmagnetische Werkstoffe Auf Ni-Fe-Basis, Report Nr. 27, VDM Metals GmbH, Werdohl, Germany, 2014.

[3] H. Rösch, Current-operated ELCBs for a.c. and pulsating d.c. fault currents, Siemens Power Engineering III 8–9 (1981) 252–255.

[4] T. Gruhn, J. Glenney, M. Savostianik, Type B ground-fault protection on adjustable frequency drives, IEEE Transactions on Industry Applications 1 (54) (2018) 934–939.

[5] R. Solleder, Allstromsensitive Fehlerstrom-Schutzeinrichtung fur Industrieanwendung, ETZ 16 (115) (1994) 896–901.

[6] G. Biegelmeier, G. Kiefer, K.-H. Krefter, Schutz in elektrischen Anlagen. Band 5: Schutzeinrichtungen, VDE-Verlag, Berlin, Offenbach, 1999.

CHAPTER 3

Types, parameters, and reliability of residual current devices

3.1 Conventional residual current devices

Residual current devices can be classified from the point of view of many various parameters. Most of them are defined and commented in the standards covering the requirements for manufacturers producing residual current devices [1–3]. The most common and important parameters, utilized during the selection of residual current devices, as well as during verification of electrical installations, are presented as follows.

3.1.1 Rated voltage V_n, number of poles and current paths

Rated voltage of a residual current device is required to be not lower than the nominal voltage of the installation. On the other hand, it should not be significantly higher (more than 20%) than the mentioned nominal voltage—it is related to RCD insulation parameters, the making and breaking capacity, as well as operation of the testing internal circuit (TEST) of the RCD. Preferred values of the rated voltage and coordination with a number of poles and current paths are presented in Table 3.1.

3.1.2 Rated residual operating current $I_{\Delta n}$

This is a value of the residual current, assigned by the manufacturer, at which the RCD shall operate under specified conditions. Recommended values mentioned in the standards are as follows:

$$0.006-0.01-0.03-0.1-0.2-0.3-0.5-1-2-3-5-10-20-30\,\text{A}$$

In practice, the most commonly used RCDs have rated residual operating current $I_{\Delta n}=30$, 100 or 300 mA. However, as it is seen before, the standards consider a wide range of this current. There is a rated current 200 mA as well as relatively high values 10, 20, and 30 A. These high values are recommended rather for differential relays cooperating with overcurrent protection devices.

Residual Current Devices
https://doi.org/10.1016/B978-0-323-89783-9.00008-4

Table 3.1 Preferred values of the rated voltage and coordination with a number of poles and current paths; according to [2].

RCD	Circuit supplying the RCD	Rated voltage of RCD in systems: 230 V, 230/400 V, 400 V	Rated voltage of RCD in systems: 120/240 V, 240 V
Single pole with two current paths	Single-phase: phase-to-neutral, phase-to-earthed middle conductor	230 V	120 V
Two pole	Single-phase: phase-to-neutral, phase-to-phase, phase-to-earthed middle conductor	230 V	120 V
	Single-phase: phase-to-phase	400 V	240 V
	Single-phase, phase-to-phase, three-wire		120/240 V
	Three-phase: four-wire (systems: 230/400 V phase-to-neutral, 230 V phase-to-phase)	230 V	
Three pole with three or four current paths	Three-phase: three-wire or four-wire (systems: 400 V, 230/400 V, 240 V)	400 V	240 V
Four pole	Three-phase, four-wire (230/400 V system)	400 V	

3.1.3 Rated residual nonoperating current $I_{\Delta no}$

This is a value of the residual current, at which and below the RCD shall not operate under specified conditions. The standard value of the rated residual nonoperating current is equal to $0.5I_{\Delta n}$.

Considering rated residual operating current and rated residual nonoperating current, it should be noted that for nonsinusoidal residual currents permissible tripping range of both these parameters is wider than for a sine wave. Table 3.2 presents the example permissible tripping ranges for various types of waveforms. Obviously, tripping ranges for nonsinusoidal currents should be referred only to the RCDs which are designed for detection of such currents. More details about the permissible tripping range of RCDs are provided in Table 12.4 (Chapter 12).

Table 3.2 Example permissible tripping ranges of RCDs for various types of waveforms according to [1,2]; $I_{\Delta n}$—rated residual operating current.

Type of waveform	Lower limit	Upper limit
Sinusoidal 50/60 Hz	$0.5I_{\Delta n}$	$I_{\Delta n}$
Pulsating direct current:		
– Without phase-angle control (half-wave); delay angle: 0°	$0.35I_{\Delta n}$	$1.4I_{\Delta n}$ [a]
– With phase-angle control; delay angle: 90°	$0.25I_{\Delta n}$	$1.4I_{\Delta n}$ [a]
135°	$0.11I_{\Delta n}$	$1.4I_{\Delta n}$ [a]
Pulsating direct current without phase-angle control + superimposed smooth DC component 6 mA	The same as for pulsating direct current without/with phase-angle control	
Smooth direct current	$0.5I_{\Delta n}$	$2I_{\Delta n}$

[a] For RCDs of $I_{\Delta n} < 30$ mA.

3.1.4 Rated current I_n

Rated current is the highest value of the current which can flow across the RCD continuously. Preferred values of the rated current are the following:

$$6-10-13-16-20-25-32-40-50-63-80-100-125-160-200-250-400-630\,\text{A}$$

The value 160 A or higher are not preferred by manufacturers and they are not likely to be found on the market.

3.1.5 Rated making and breaking capacity I_m

Residual current devices are obliged to make and break the current of value at least $10I_n$ or 500 A, whichever is the greater. This ability is required for RCDs without backup protection by overcurrent protection devices. During the test, current of value equal to 500 A flows through the main terminals of the RCD and then residual current $10I_{\Delta n}$ is forced (Fig. 3.1A). Both poles of the two-pole RCD are breaking current equal to 500 A. Proper operation and possible damage of the RCD after the test are verified.

3.1.6 Rated residual making and breaking capacity $I_{\Delta m}$

Requirements are similar to these mentioned prior ($10I_n$ or 500 A, whichever is the greater). However, types of tests I_m vs. $I_{\Delta m}$ are different. Fig. 3.1B

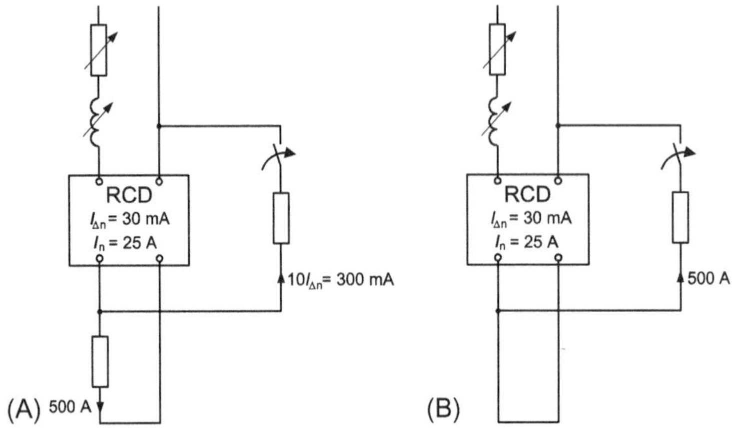

Fig. 3.1 Testing an RCD ($I_{\Delta n}=30\,\text{mA}$, $I_n=25\,\text{A}$): (A) rated making and breaking capacity I_m, (B) rated residual making and breaking capacity $I_{\Delta m}$.

presents the principle of this test. A current of value 500 A flows as a residual current. Only one pole of the RCD is breaking current equal to 500 A (worse breaking conditions for the RCD than during I_m test). Proper operation and possible damage to the RCD after the test are verified.

3.1.7 Rated conditional short-circuit current I_{nc}

This is a value of the prospective current (the AC component) which RCD, backup protected by a suitable overcurrent protective device, can withstand under specified conditions of use and behavior.

From this point of view two types of RCDs are distinguished:

(a) RCCB—residual current operated circuit-breaker without integral overcurrent protection; this type of RCD is not designed to perform the functions of protection against overloads and short-circuits. RCCB is required to be backup protected, and example marking with regard to the backup protection is presented in Table 3.3.

Preferred values of the I_{nc} are as follows:

$$3\,\text{kA} - 4.5\,\text{kA} - 6\,\text{kA} - 10\,\text{kA} - 20\,\text{kA} - 25\,\text{kA}$$

(b) RCBO—residual current operated circuit-breakers with integral overcurrent protection; this type of RCD is composed of a residual current device and a circuit-breaker. Thus it is able to protect circuits against overloads and short-circuits. It is marked, e.g., C16 (the type of tripping characteristic) and [6000] (rated conditional short-circuit current equal to 6 kA) as a typical circuit-breaker.

Table 3.3 Marking of RCCBs with regard to their backup protection.

Graphical symbol	Description
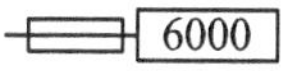	Rated conditional short-circuit current equal to 6 kA, provided that backup protection by gG fuse of a rated current not exceeding 63 A is used
	Rated conditional short-circuit current equal to 6 kA, provided that backup protection by gG fuse of a rated current not exceeding 100 A is used
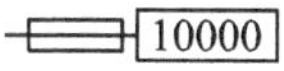	Rated conditional short-circuit current equal to 10 kA, provided that backup protection by gG fuse of a rated current not exceeding 63 A is used

3.1.8 Rated conditional residual short-circuit current $I_{\Delta c}$

Value of a residual prospective current (the AC component) which RCD, backup protected by a suitable overcurrent protective device, can withstand under specified conditions of use and behavior.

Requirements, standard, and preferred values for this current are similar to those presented in case of conditional short-circuit current I_{nc}.

3.1.9 Operating characteristics in terms of the residual current waveform shape

Proper selection of an RCD with regard to waveform shape of the residual current is extremely important, because in some cases the RCD may not react at all to the residual current significantly exceeding the rated residual operating current [4–10]. Table 3.4 presents types of RCDs with a division into the abilities of detection of particular waveform shape.

AC residual current devices are able to detect sinusoidal (50/60 Hz) residual currents only. This is the simplest type of RCDs. Due to the widespread use of electronic current-using equipment, which produces direct or mixed-frequency earth fault currents, in some countries, AC-type RCDs are not allowed to be installed. For example, in Denmark since the year 1994, newly installed RCDs have to be at least A-type [12]. Application of at least A-type RCDs should be nowadays the common practice. They are able to detect pulsating earth faults current which may be produced by power supply units included in electronic equipment. However, it should be underlined that A-type RCDs are not sufficient in circuits, in which mixed-frequency/composite earth fault current flows (e.g., circuits supplying equipment containing PWM variable-speed drives).

Table 3.4 Operating characteristics of RCDs in terms of the residual current waveform shape—according to [1,2,11].

Alphanumerical and graphical symbol	Waveform shape for which tripping of RCDs is ensured
AC or 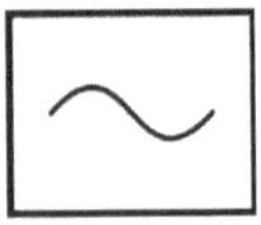	– Sinusoidal alternating current (usually 50/60 Hz)
A or 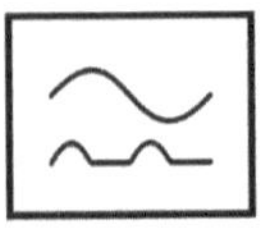	– Sinusoidal alternating current (usually 50/60 Hz), – Pulsating direct currents, – Pulsating direct currents with smooth direct component not exceeding 6 mA, (also for specified current delay angles)
F  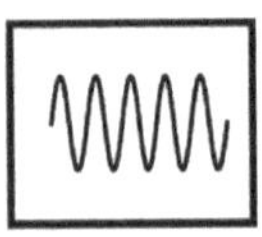 or 	– Residual currents as for A-type RCDs, – Pulsating direct currents with smooth direct component not exceeding 10 mA, – Composite residual current (mixed frequencies), supplied from single phase
B 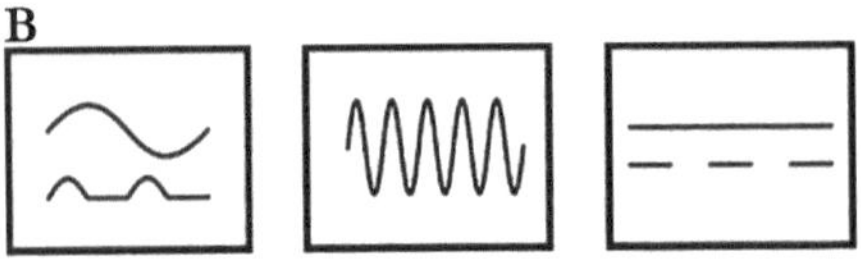 or	– Residual currents as for F-type RCDs, – Residual sinusoidal alternating currents up to 1000 Hz, – Residual alternating currents superimposed on a smooth direct current of 0.4 times the rated residual operating current, – Residual pulsating direct currents superimposed on a

Table 3.4 Operating characteristics of RCDs in terms of the residual current waveform shape—according to [1,2,11]—cont'd

Alphanumerical and graphical symbol	Waveform shape for which tripping of RCDs is ensured
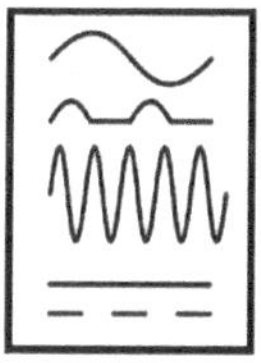	smooth direct current of 0.4 times the rated residual operating current or 10 mA, whichever is the highest value, – Residual direct currents which may result from rectifying circuits, i.e.: – two-pulse bridge connection line to line for 2-, 3-, and 4-pole devices, – three-pulse star connection or six-pulse bridge connection for 3- and 4-pole devices, – Residual smooth direct currents

For such types of currents, F-type RCDs are provided. Table 3.5 presents parameters of current which is utilized for testing F-type RCD. This current is composed of a rated frequency component (usually 50/60 Hz), a 1 kHz component, and a 10 Hz component. This test waveform reflects a mixed-frequency earth fault current in single-phase circuits containing variable-speed drives. Tripping threshold for such a waveform is required to be within the range $(0.5–1.4)I_{\Delta n}$.

In case of a three-phase circuit containing variable-speed drives, a B-type RCD shall be installed, because in such a circuit smooth direct earth fault current may flow. B-type RCDs are obviously suitable for the detection of mixed-frequency currents.

Standards IEC 60755 [1], IEC 61008-1 [2], and IEC 62423 [11] present diagrams according to which RCDs should be verified in terms of their proper operation under defined waveform. Fig. 3.2 presents a diagram for verification of the RCD operation in case of AC sinusoidal waveform (for AC-type, A-type, and F-type RCDs: under network frequency, usually 50/60 Hz; for B-type RCDs: for frequency up to 1000 Hz). The resistor of the variable resistance R_s and generator G_s enables for control of the residual current. When the correct operation of the RCD under a steady increase of

Table 3.5 Parameters of the test current of F-type RCDs and their operating current ranges according to [11].

Components of the test current (rms)			Composite test current (rms)	
Rated frequency	**1 kHz**	**10 Hz**	**Starting value**	**Tripping range**
$0.138I_{\Delta n}$	$0.138I_{\Delta n}$	$0.035I_{\Delta n}$	$0.2I_{\Delta n}$	$(0.5–1.4)I_{\Delta n}$

$I_{\Delta n}$—rated residual operating current at the rated frequency.

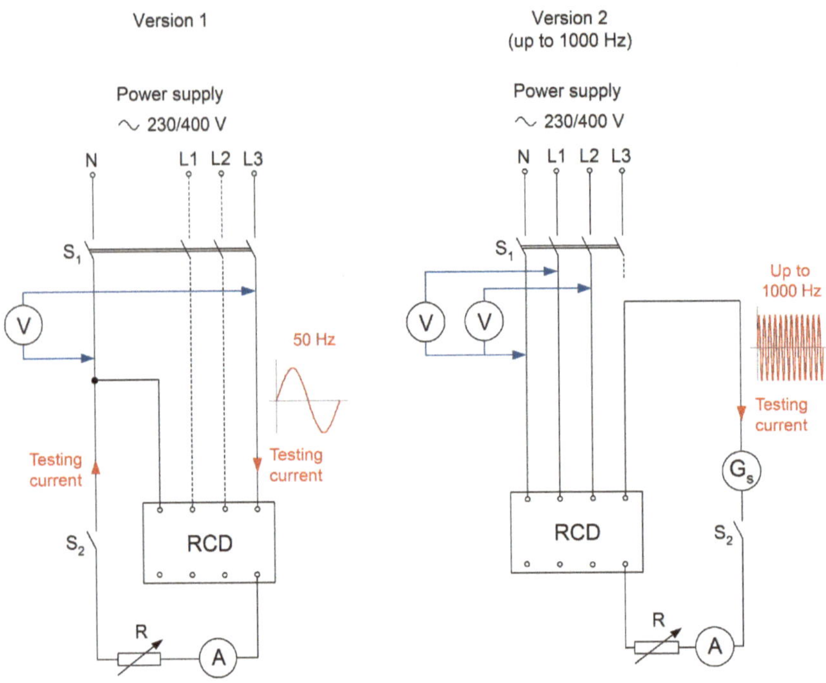

Fig. 3.2 Test circuit for the verification of the RCD operation in the case of residual sinusoidal alternating current 50 Hz or residual sinusoidal alternating current up to 1000 Hz (for B-type RCDs) (S_1—all-pole switch, S_2—single-pole switch, R—variable-resistance resistor, A—ammeter, V—voltmeter, G_s—generator for producing sinusoidal waveforms (in the case of B-type RCD testing, the frequency range is up to 1000 Hz), RCD—residual current device).

the residual current is verified, the switches S_1 and S_2 are in a closed position and the residual current is steadily increased by changing the value of the resistance R(var). The ammeter A is used to check the tripping current value. According to this diagram, a verification of the RCD operation in the case of the sudden appearance of residual current can be verified as well. After establishing the relevant value of the residual current (e.g., characteristic values $I_{\Delta n}$, $2I_{\Delta n}$, and $5I_{\Delta n}$ as specified in Fig. 3.11), switches S_1 and S_2 are closed consecutively. The tested RCD shall trip.

The verification of the RCD operation in the case of residual pulsating direct currents should be performed according to the diagram depicted in Fig. 3.3. Thyristors enable to force pulsating direct currents of various current delay angles. The ammeter is required to be suitable for the measurement of the rms value of nonsinusoidal waveforms.

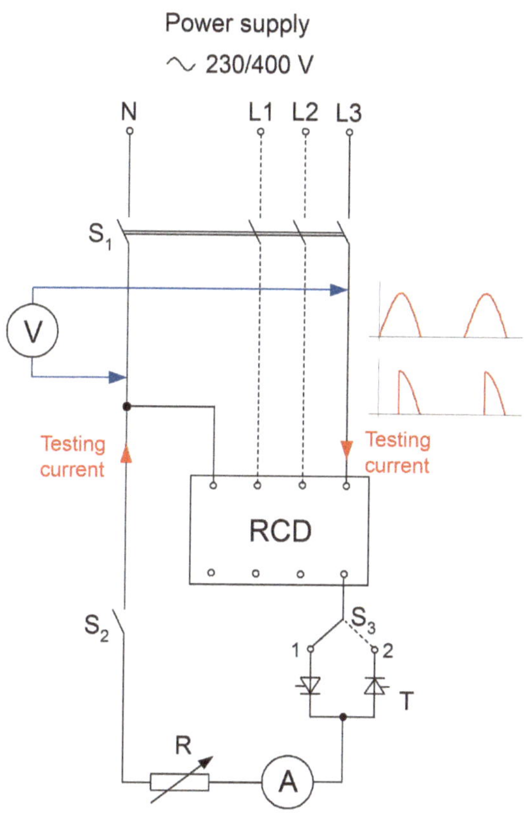

Fig. 3.3 Test circuit for the verification of the RCD operation in the case of residual pulsating direct currents (also for specified current delay angles) (S_1—all-pole switch, S_2—single-pole switch, S_3—two-way switch, R—variable-resistance resistor, A—ammeter (measuring true rms values), V—voltmeter, T—thyristors, RCD residual current device).

The widest spectrum of the verification of RCD operation under DC residual currents is provided for B-type RCDs. In addition to the previously mentioned tests, RCDs are tested under:

- a residual alternating current superimposed on a smooth direct current (Fig. 3.4); during this test, a smooth DC current is applied (from a separate power source; a DC current is measured by ammeter A_1) along with the supply with the AC power network (an AC current is measured by ammeter A_2),
- a residual current composed of a pulsating direct component and a smooth direct component (Fig. 3.5); the pulsating DC component may have a specified current delay angle,
- a residual pulsating direct current which may result from rectifying circuits supplied from two phases (Fig. 3.6),

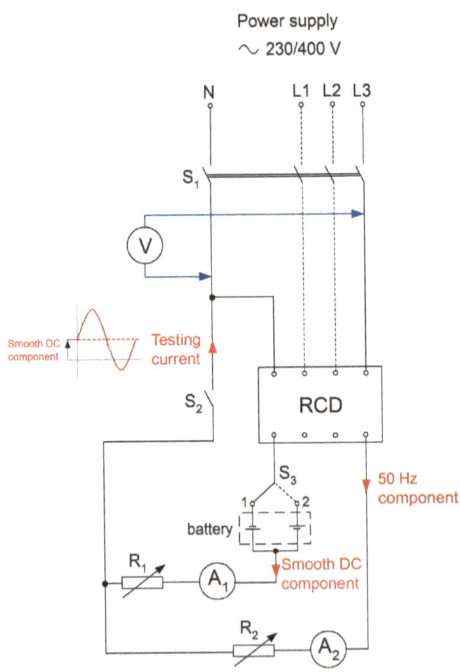

Fig. 3.4 Test circuit for the verification of the 2-, 3-, and 4-pole B-type RCD operation in the case of a residual alternating current superimposed on a smooth direct current (S_1—all-pole switch, S_2—single-pole switch, S_3—two-way switch, R_1, R_1—variable-resistance resistors, A_1—ammeter for DC currents, A_2—ammeter for AC currents, V—voltmeter, RCD—residual current device).

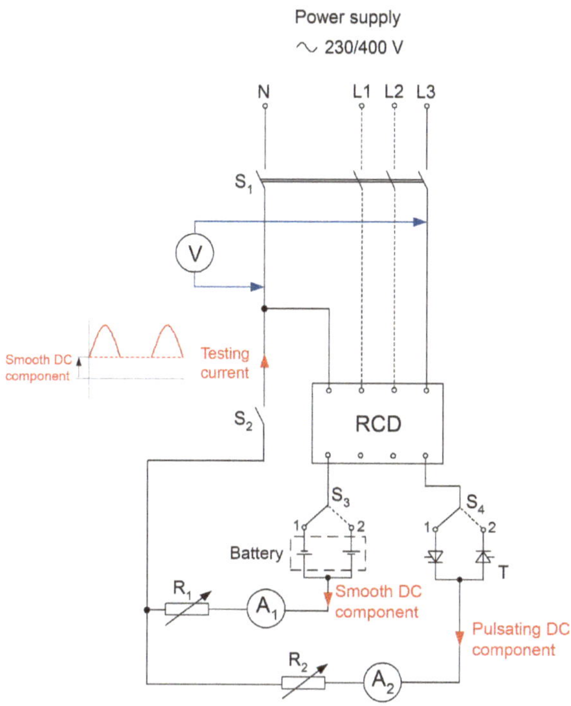

Fig. 3.5 Test circuit for the verification of the 2-, 3-, and 4-pole B-type RCD operation in the case of a residual pulsating direct current (also for specified current delay angles) superimposed on a smooth direct current (S_1—all-pole switch, S_2—single-pole switch, S_3, S_4—two-way switches, R_1, R_1—variable-resistance resistors, A_1—ammeter for DC currents, A_2—ammeter (measuring true rms values), V—voltmeter, T—thyristors, RCD—residual current device).

- a residual pulsating direct current which may result from rectifying circuits supplied from three phases (Fig. 3.7),
- a residual smooth direct current (Fig. 3.8).

A specific test is provided for F-type RCDs (also for B-type RCDs). In addition to the verification of the operation under a sinusoidal current (Fig. 3.2) and pulsating DC current (Fig. 3.3), they are tested under a mixed waveform as specified in Table 3.5 (10 Hz + 50 Hz + 1000 Hz). For such a test, an arbitrary generator for producing mixed-frequency waveforms should be used (Fig. 3.9).

The aforementioned diagrams for the RCD testing have been assumed as a reference for the laboratory stand arrangement during the tests performed

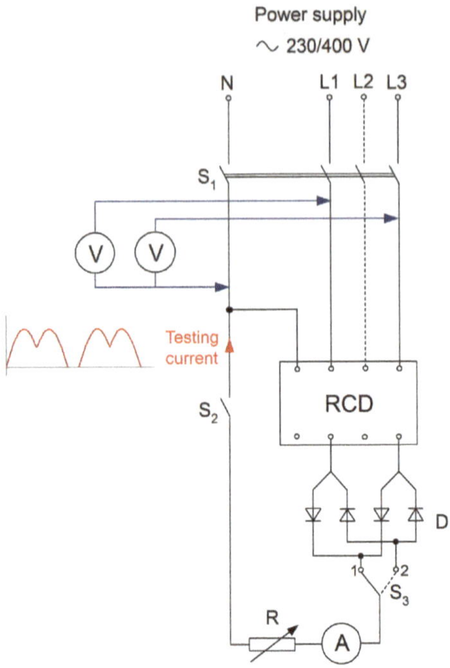

Fig. 3.6 Test circuit for the verification of the 2-, 3-, and 4-pole B-type RCD operation in the case of a residual pulsating direct current which may result from rectifying circuits supplied from two phases (S_1—all-pole switch, S_2—single-pole switch, S_3—two-way switch, R—variable-resistance resistor, A—ammeter (measuring true rms values), V—voltmeter, D—diodes, RCD—residual current device).

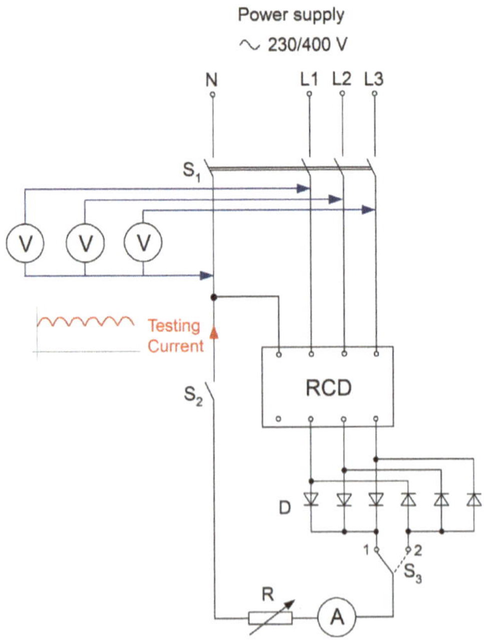

Fig. 3.7 Test circuit for the verification of the 2-, 3-, and 4-pole B-type RCD operation in the case of a residual pulsating direct current which may result from rectifying circuits supplied from three phases (S_1—all-pole switch, S_2—single-pole switch, S_3—two-way switch, R—variable-resistance resistor, A—ammeter (measuring true rms values), V—voltmeter, D—diodes, RCD—residual current device).

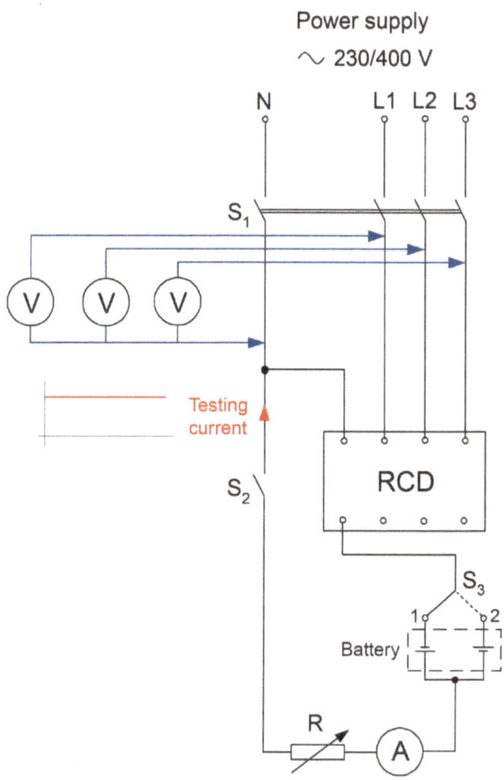

Fig. 3.8 Test circuit for the verification of B-type RCD operation in the case of a residual smooth direct current (S_1—all-pole switch, S_2—single-pole switch, S_3—two-way switch, R—variable-resistance resistor, A—ammeter for DC currents, V—voltmeter, RCD—residual current device).

by the author. Results of the author's tests of the RCD tripping are included mainly in Chapter 8.

In addition to the above-presented types of RCDs, B+ type RCDs are also available, produced according to German standard DIN VDE 0664-400. This type of RCD has an extended ability for detection of residual sinusoidal alternating currents of higher frequency—it is suitable for residual currents up to 20 kHz [13,14]. Fig. 3.10 presents tripping characteristic of 30 mA RCDs, as a function of residual current frequency, according to the manufacturer Doepke.

It is worth mentioning that the tripping threshold of this characteristic changes with frequency rising. From frequency 1000 Hz, the upper threshold is 420 mA (14 × 30 mA). It is the equivalent level of the threshold of ventricular fibrillation [15] for frequencies 1000 Hz and above, indicated in report [16]. Currents of higher frequency are less dangerous for persons than currents of frequency 50/60 Hz.

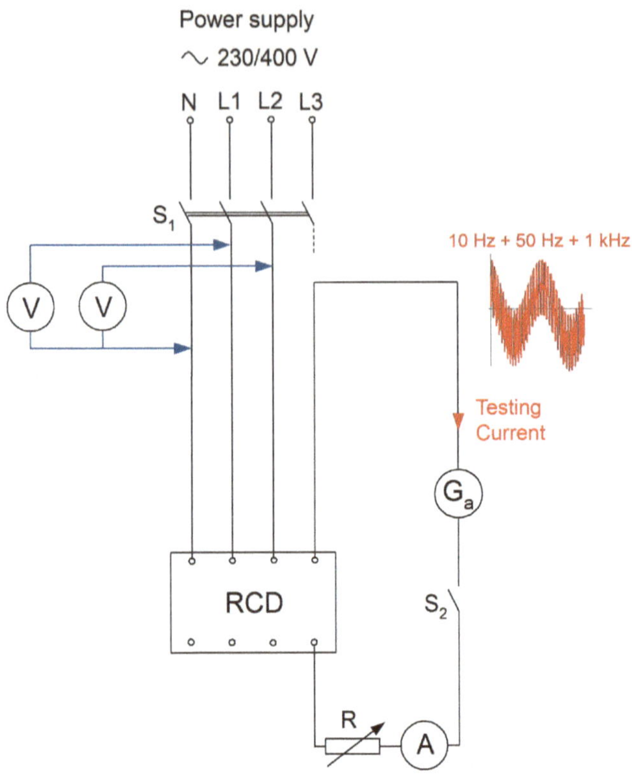

Fig. 3.9 Test circuit for the verification of the RCD operation (for F-type and B-type RCDs) in the case of a residual sinusoidal alternating current composed of multifrequency components resulting from a single-phase supply of motor control equipment (for motor speed control) (S_1—all-pole switch, S_2—single-pole switch, R—variable-resistance resistor, A—ammeter (measuring true rms values), V—voltmeter, G_a—arbitrary generator for producing mixed-frequency waveforms (10 Hz + 50 Hz + 1 kHz). RCD—residual current device).

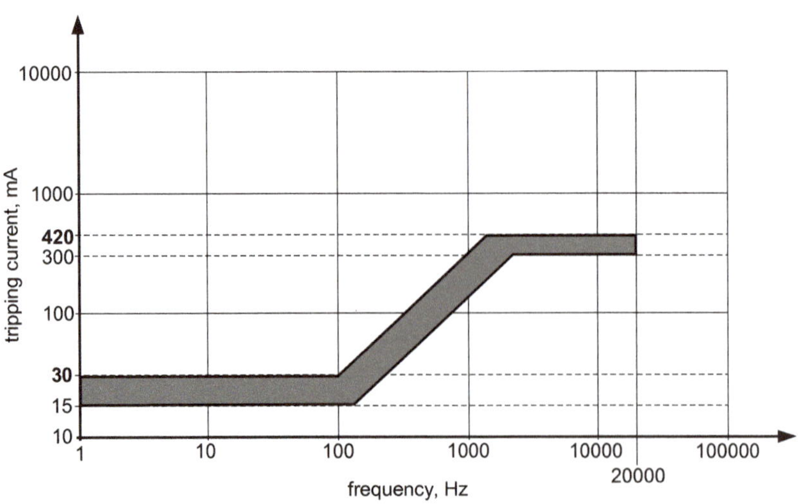

Fig. 3.10 Example tripping characteristic of 30 mA RCDs (B+ type) as a function of the frequency of the residual current.

3.1.10 Delay in tripping

From this point of view, following are the types of RCDs:

- no-delayed—without specified delay time, they are not specially marked,
- short-time delayed—with the guaranteed delay time equal to 10 ms, marked [G], VSK, KV, or HI; such devices are dedicated to the circuits with electronic equipment, which are characterized by transient leakage currents of high value; note that standard IEC 62423 [11] requires that all F-type and B-type RCDs are able to withstand inrush residual currents with a maximum duration of 10 ms,
- delayed—with the guaranteed delay time equal to 40 ms, marked [S]; such RCDs are installed in distribution circuits and they are intended to selective coordination with no-delayed and short-time delayed RCDs.

There are also residual current relays, composed of an external current transformer and a relay unit. These relays enable for adjusting both residual tripping current, e.g.:

$$0.03 - 0.1 - 0.3 - 1 - 3 - 5 - 10\,\mathrm{A}$$

and time delay, e.g.:

$$0.06 - 0.1 - 0.3 - 1 - 5\,\mathrm{s}$$

Fig. 3.11 presents the time–current tripping characteristics of the example RCDs. In every tripping characteristic, normative points presenting maximum permissible break times and minimum nonoperating times, for particular residual currents, are indicated. For example, in case of no-delayed and short-time delayed RCDs, for residual current I_Δ equal to the rated residual operating current $I_{\Delta n}$, the maximum permissible break time is equal to 0.3 s. In case of delayed RCDs (S-type), this time is 0.5 s.

3.1.11 Permissible ambient temperature range

Tripping mechanism of residual current devices is sensitive to various external influences, in particular, humidity and temperature. Due to low (or very high) ambient temperature, the tripping threshold of RCDs may deviate from the rated parameter, therefore it is necessary to specify the permissible ambient temperature range. The following temperature ranges are usually met:

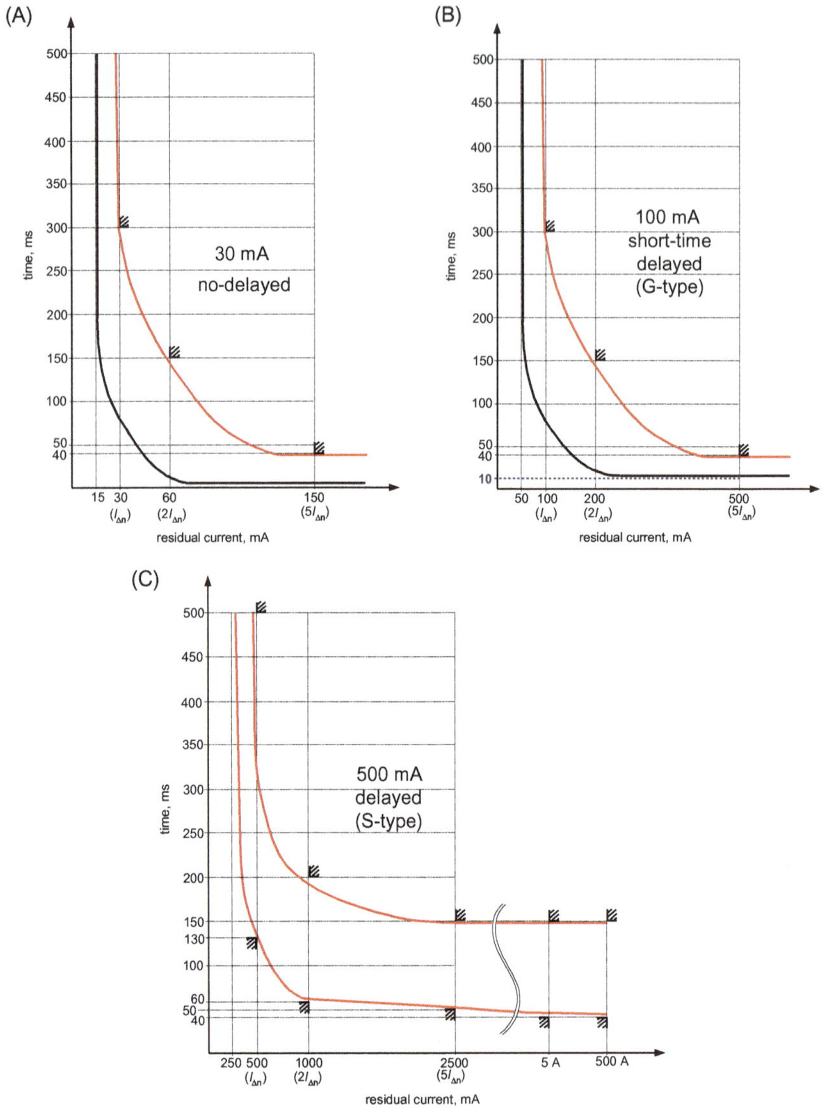

Fig. 3.11 Time–current characteristic for various types of RCDs (for sinusoidal waveform) taking into account delay in tripping: (A) 30 mA no-delayed, (B) 100 mA short-time delayed, (C) 500 mA delayed.

- from 5 °C to +40 °C—a standard range, such RCDs may be installed only indoors,
- from – 25 °C to +40 °C—frost-proof RCDs, marked , such RCDs may be installed outdoors.

3.2 Special residual current devices and their functions

There are special constructions and types of RCDs from the point of view of their technical solution or purpose of the application. Some of them are met relatively often in electrical installations; others, very rarely (only in special applications); and others are still just a concept. Selected special types of RCDs, which may be of more interest to readers, are described as follows.

3.2.1 RCD with the emergency STOP function

Safe utilization of some devices and systems, especially in the industry (e.g., people working with moving/rotating machines), requires the use of emergency stopping. Such a stopping is triggered by the emergency stopping buttons—pressing the button breaks the auxiliary circuit and, therefore, the main circuit supplying a moving machine is disconnected/stopped.

The emergency STOP function can be integrated with a special type of RCD (Fig. 3.12). The current transformer of such an RCD has three secondary windings:

- a winding (2) for supplying the electromechanical relay (1), as in a classic RCD,
- two auxiliary windings (4-1 and 4-2) with the same number of turns but these turns (4–1 vs. 4–2) are performed in the opposite directions; the turns are supplied with equal voltage.

In such an RCD, the current of the same value is flowing through these two auxiliary windings. The current value is a result of the supply voltage and resistance R as well as two resistances $R/2+R/2$. When the emergency STOP button is not pressed, magnetic fluxes generated by two auxiliary windings compensate each other; hence, the RCD is closed. One of the auxiliary winding (4-1) is connected with the external circuit having emergency STOP buttons. Pressing any of the emergency STOP buttons shown in Fig. 3.12 makes that the external circuit is opened, no current flows in this circuit, and no magnetic flux is generated in the respective auxiliary winding (4-1). In consequence, the magnetic flux produced by the second auxiliary

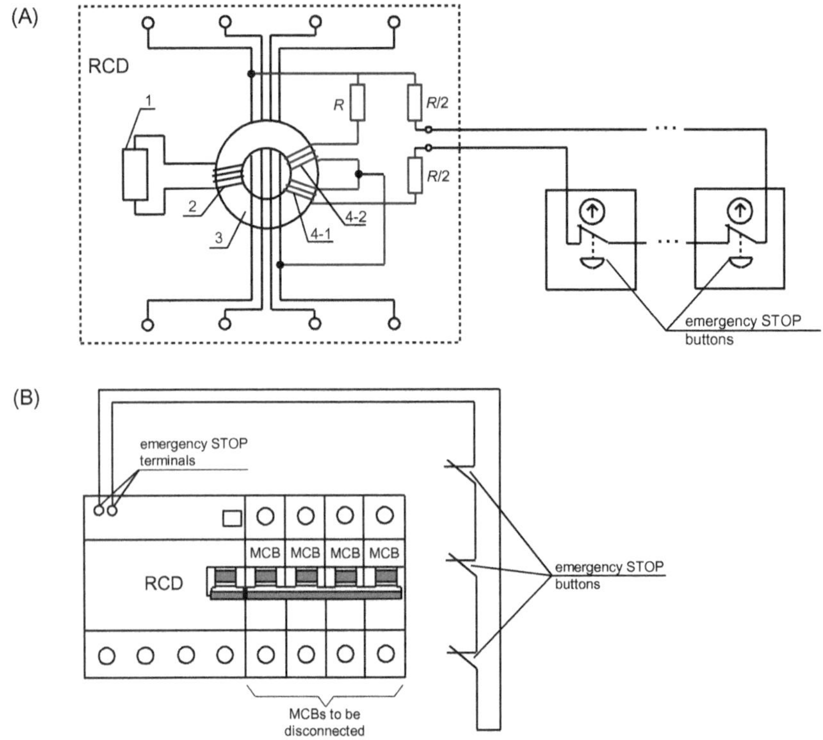

Fig. 3.12 RCD with the emergency STOP function: (A) internal structure, (B) integration with external circuit-breakers (MCBs) and circuits to be disconnected; according to [17]; 1—electromechanical relay, 2—secondary winding connected with the relay, 3—iron core of the current transformer, 4–1, 4–2—auxiliary windings for emergency STOP purpose, *R*, *R*/2—reference resistors in the winding for emergency STOP purpose.

winding (4-2) is not compensated, which makes tripping of the relay (1) and, hence, the RCD as well. Compared to other solutions with active safety, the use of the RCD as an emergency STOP device has a definite advantage—it is not negatively influenced by voltage dips/sags. The aforementioned type of RCD is recommended in circuits of escalators, elevators, controlled gates, automatic car washes, production belts, belt conveyors, and machine tools such as lathes and milling machines.

3.2.2 RCD with the short-circuit system for disconnected circuits having the remaining voltage

The concept of the safety system based on an RCD refers to circuits with any energy storage device, e.g., capacitors or motors, which still rotates after switching off the circuit. In the case of the basic insulation fault in a motor circuit, after disconnection of supply by a classic RCD or circuit-breaker, the risk of electric shock still exists due to the residual voltage remaining between the motor's live terminals and in consequence between these terminals and earth. Such a risk can be eliminated by short-circuiting live terminals/conductors immediately after the circuit has been disconnected by the RCD. In 230/400 V TN systems, in many final circuits, the maximum permissible disconnection time in case of earth fault is equal to 0.4 s. In the TT system, it is 0.2 s. Low-voltage motors for household and similar uses may produce at their terminals dangerous voltage higher than 50 V (after disconnecting from the power supply) within a time much longer than 0.5 s. By the use of the RCD with the in-built short-circuit system presented in Fig. 3.13, this time can be shortened to be below 0.1 s. Immediately, after opening the main contacts of the RCD, the integrated short-circuit system (3) connects all live conductors together to discharge the energy stored in a capacitor or motor.

3.2.3 RCD with the reclosing system

The susceptibility of RCDs to unwanted tripping resulting from atmospheric overvoltages or switching processes as well as from other disturbances in the supply network is a serious drawback in terms of the reliability of supply of current-using equipment requiring uninterrupted operation (e.g., water pumps, air conditioning, telecommunication/GSM devices, remote sensing, data transmission). If the reclosing of such equipment has to be done manually by service staff, it means a long time without supply and in consequence, may generate high additional costs. A remedy to this problem, called *restart RCD*, appeared at the end of the 1990s of the twentieth century. It is an auto-reclosing device that may be either a separate apparatus mechanically connected to the RCD [18] or a system in-built into the RCD [19]. The restart device may serve a triple or fivefold auto-reclosing cycle.

Fig. 3.14 presents the auto-reclosing module connected to the RCD as well as the module operation algorithm. This device serves a fivefold auto-reclosing cycle. After detection of a residual current I_Δ, the RCD opens the protected circuit and, therefore, makes a reaction (to open position) of the coupled module. The first reclosing appears after 20 s. If the RCD opens the circuit again, the next auto-reclosing by the module will occur within 30 s.

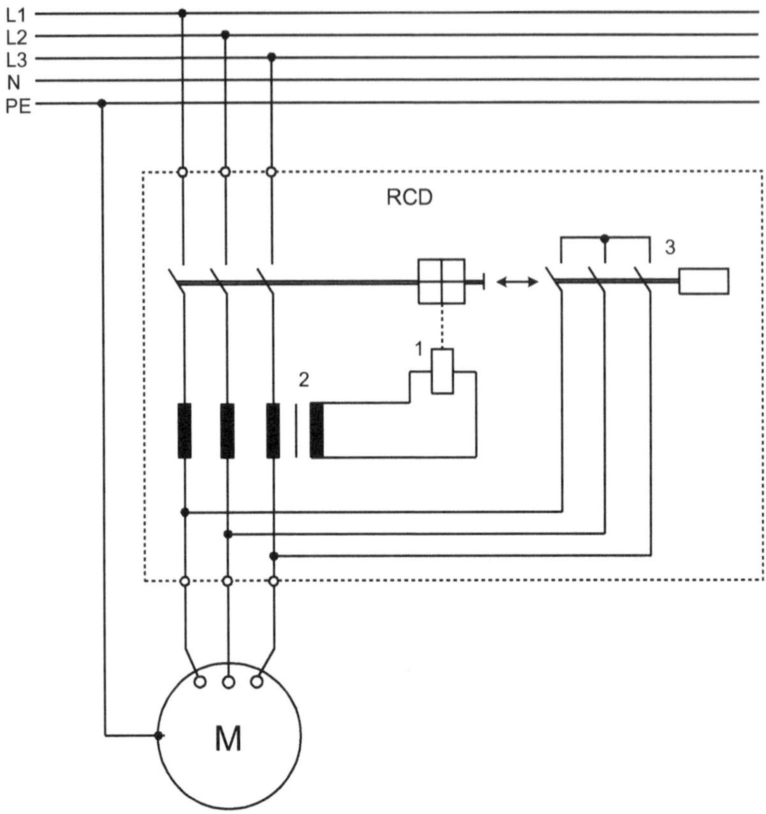

Fig. 3.13 RCD with the short-circuit system for disconnected circuits: 1—electromechanical relay, 2—current transformer, 3—integrated short-circuit system to discharge the energy stored in loads after their disconnection from the supply, M—motor.

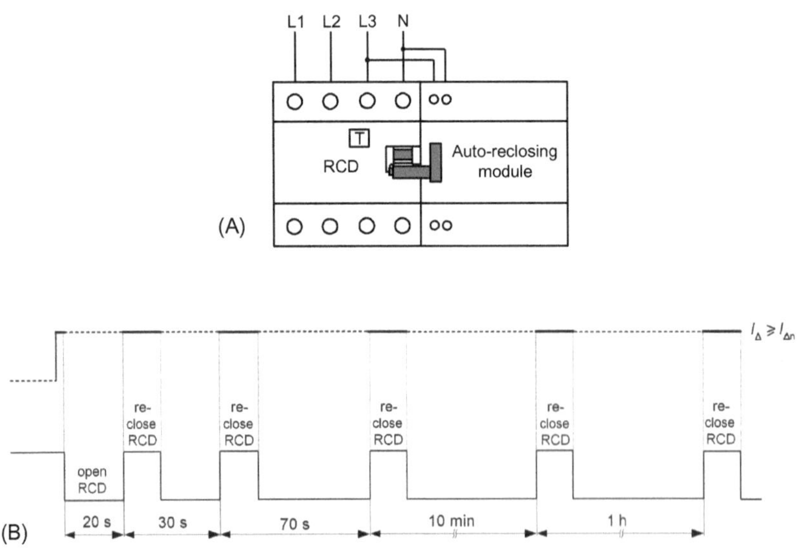

Fig. 3.14 RCD coupled with the auto-reclosing module: (A) a simplified view of the coupled devices, (B) operation algorithm; according to [18].

Each subsequent period between consecutive auto-reclosings is longer—the final period (between 4th and 5th reclosing) is equal to 1 h.

The auto-reclosing of the RCD is desired from the point of view of the reliability of supply but can be controversial in terms of the protection against electric shock—a person touching faulty current-using equipment or touching live terminal (only the 30 mA RCD, tripping immediately, is effective) may be electrocuted several times. Even in a maintenance-free facility (e.g., GSM substation), repair or modernization works are carried out from time to time, during which service staff use electrical equipment, including hand tools. It is not allowed to carelessly expose the staff to the repeated body current flow. With the aforementioned in mind, the solution presented in Fig. 3.15, having an in-built auto-reclosing module, offers an additional function—the verification of the condition of the protected circuit.

In this solution, automatic reclosing of the RCD is possible only after a positive result of the protected circuit test. For this purpose, the device is equipped with an electronic circuit that monitors the protected circuit. The system for circuit testing is powered by the network voltage and is activated after the RCD tripping. The circuit testing is performed with a pulsating DC current of a value not exceeding 0.1 mA (it is safe for humans and animals). It roughly verifies the resistance of the circuit to earth, which in an undamaged circuit equals the insulation resistance to earth of live conductors (usually it is many MΩ). The measured resistance lower than a few kΩ indicates that the insulation to earth is damaged or a human touches a live part in the circuit. If so, the system stops reclosing. Otherwise (a positive result of the test), the first reclosing is possible—it occurs after 90 s. Max three reclosings are possible within the specified time. The different types of these devices can be found in [20].

The issue of the automatic reclosing of RCDs has been normalized—for this purpose international standard IEC 63024 [21] classifies Automatic Reclosing Devices (ARDs) and specifies requirements for their performance and tests.

3.2.4 Portable RCD

The portable high-sensitivity residual current device (PRCD) of a rated voltage of 230 V and rated current $I_n = 16$ A can provide additional protection of electric shock in the case of indirect contact in single-phase AC final circuits. PRCDs have an in-built high-sensitivity residual current device

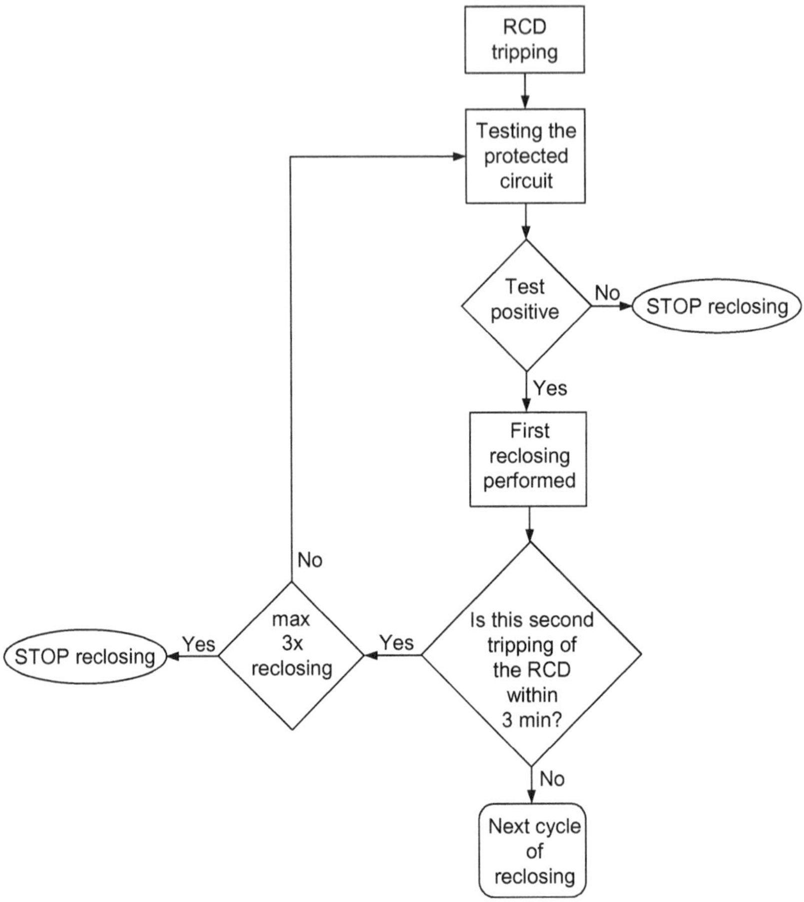

Fig. 3.15 Operation algorithm of the RCD with the reclosing system and verification of the protected circuit; according to [19].

($I_{\Delta n} \leq 30\,\text{mA}$) and may have an overvoltage as well as undervoltage protective function. PRCDs are designed according to the standard [22] and they are to be installed:

- in single and branch extension cords,
- in portable power lines supplying cable drums and small switchboards in construction/demolition sites,
- in the plugs of portable cables of low-power receivers.

PRCDs are suitable to ensure the safety of mobile devices use, especially hand tools, in conditions with a particular risk of electric shock, in damp, wet, or dusty rooms and in the open air: on construction and demolition sites, in agriculture and horticulture premises, at camps to supply installations in tents and vehicles.

PRCDs are expected to detect not only residual current but also to react to some other faults or mistakenly incorrect connections, mainly: broken PE conductor in the protected circuit and the appearance of dangerous voltage on PE conductor. Verification of 24,000 installations in Germany in 1994 showed that as much as 24% of them had broken PE conductor. It proves that such additional functions of the PRCDs are strongly recommended.

Unfortunately, early PRCDs have been constructed with the nondisconnected PE conductor going outside the current transformer of the PRCD. In such a solution, in the case of an insulation fault in the circuit protected by the PRCD (Fig. 3.16A) as well as in the case of an insulation fault in the external circuit (resulting in voltage on the PE conductor in the circuit with the PRCD—Fig. 3.16B), the risk of electric shock still exists because the PRCD does not react. A construction of the PRCD presented

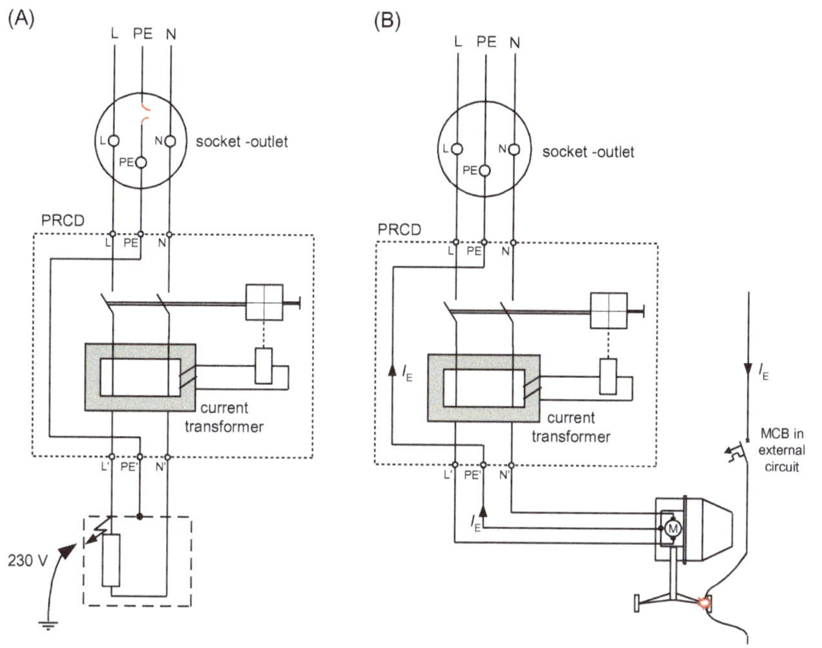

Fig. 3.16 No reaction of the PRCD with nondisconnected PE conductor going outside the current transformer, in the case of: (A) an insulation fault and broken continuity of the PE conductor, (B) occurrence of the earth current I_E in the PE conductor, due to an insulation fault in the external circuit.

in Fig. 3.17 has the PE conductor going through the current transformer but in the opposite direction and with more number of turns than L and N conductors. It enables for detection of the voltage on the PE conductor in the case of the insulation fault in the external circuit. However, after disconnection of supply by the PRCD, the current-using equipment (a concrete mixer in Fig. 3.17) still has voltage on the exposed-conductive-parts. Moreover, the circuit-breaker (MCB) in the external circuit may not disconnect the supply because the disconnected PE conductor breaks the continuity of the earth current path.

In order to eliminate the aforementioned weak points of the PRCDs, it has been searched for a solution that would disconnect the PE conductor in some cases but in the others, this conductor would have the continuity. Such a solution is included in a Portable Residual Current Device—Safety (PRCD-S) having an electronic system for analyzing the condition of the protected circuit. All conductors can be disconnected but the PE conductor does not go through the current transformer (Fig. 3.18). The PRCD-S:

- does not allow to switch on the protected circuit if the conductors N or PE are broken,

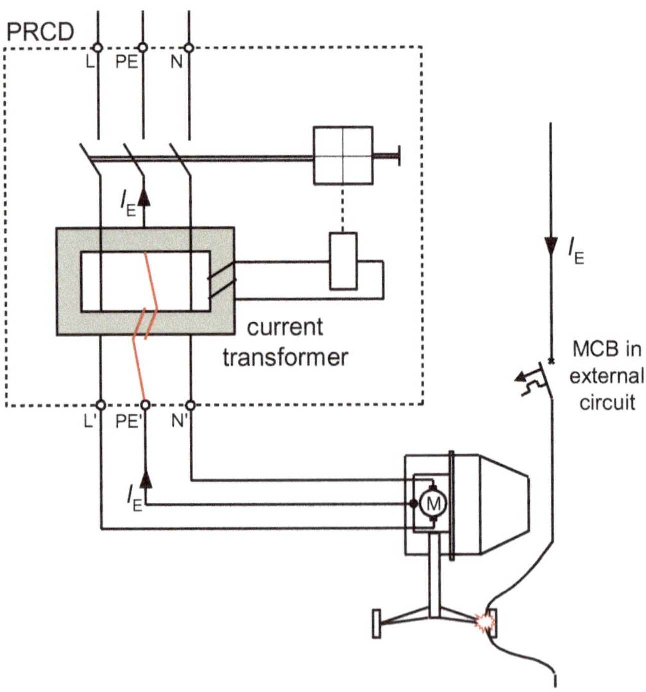

Fig. 3.17 PRCD with the PE conductor going in the opposite direction through the current transformer; it enables for detection of the earth current I_E from the external circuit.

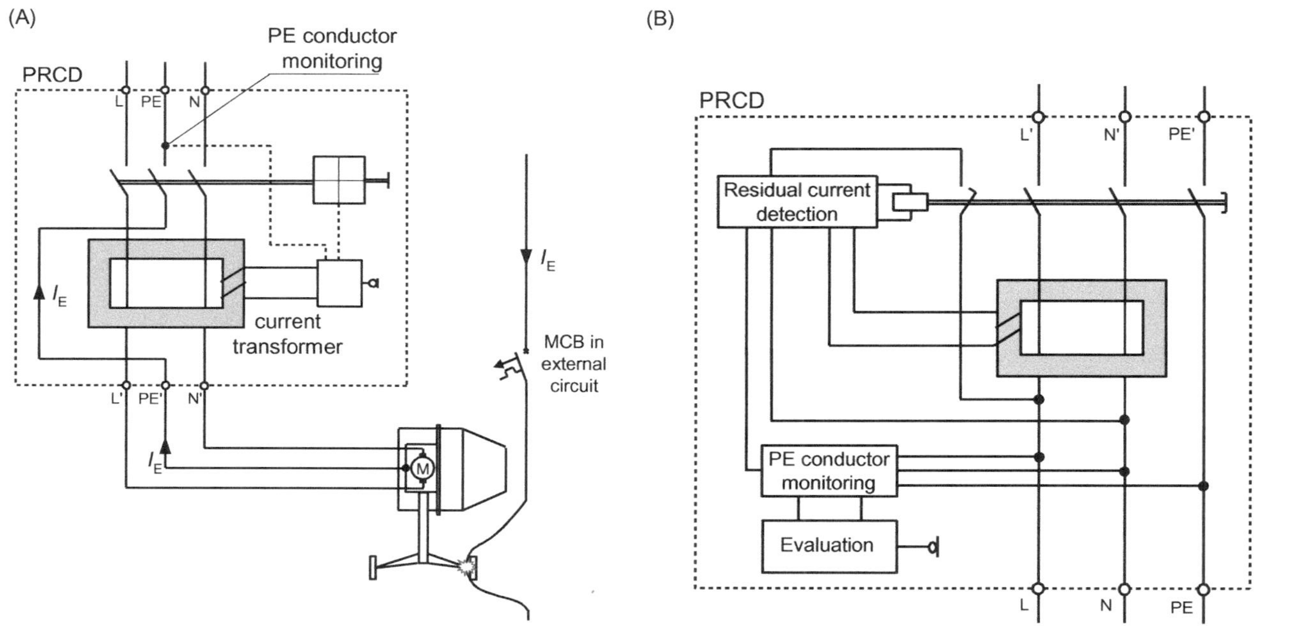

Fig. 3.18 PRCD-S with disconnected PE conductor going outside the current transformer, and monitoring of the PE conductor: (A) general structure, (B) detailed structure; according to [23].

- does not allow to switch on the protected circuit if the conductors L and PE are swapped,
- does not allow to switch on the protected circuit if the PE conductor is under dangerous voltage,
- disconnects the supply by opening all conductors (including PE conductor) in the case the upstream PE conductor is broken,
- disconnects the supply by opening the conductors L and N (the PE is not disconnected) in the case of occurrence of voltage from the external circuit (the earth current is able to flow through the PE conductor and it enables for the reaction of the MCB in the external circuit—Fig. 3.18A).

The example PRCD with complex protective functions is presented in Fig. 3.19. It is called PRCD-S+. After connecting the device to the

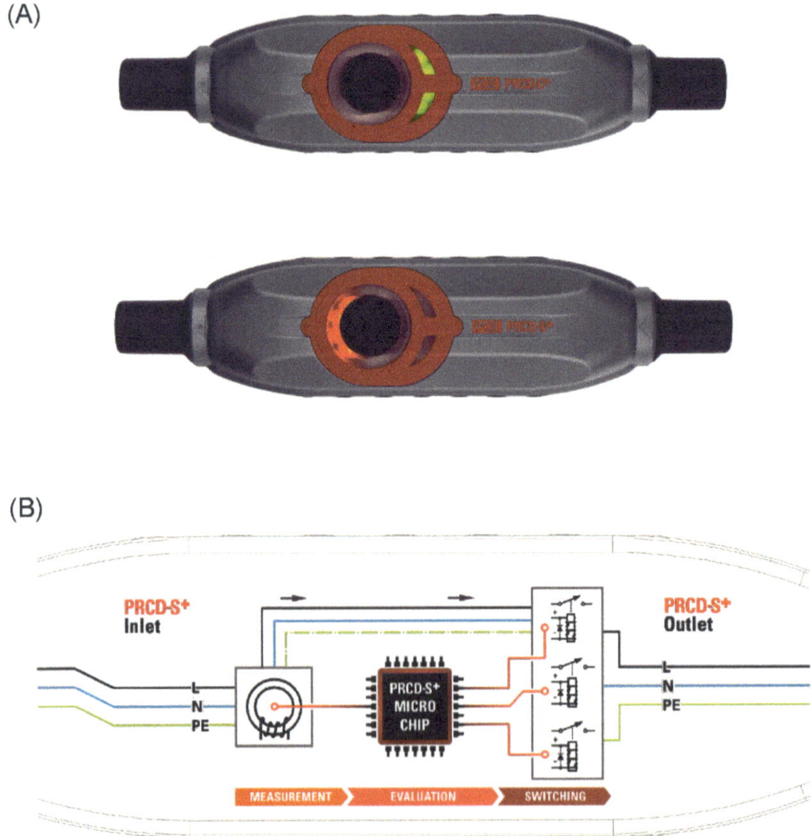

Fig. 3.19 PRCD-S+ with complex protective function: (A) an external view, (B) internal structure (manufacturer PC Electric) [24].

installation, self-testing is performed. To use the PRCD-S+, a user should press the button. If no problem occurs in the circuit, the LED indicator lights in green color. The ON/OFF button has a safety function—it verifies the potential of the user to earth. Therefore the user should press the button with bare hands. If the result of the test is negative, the switching on the circuit is not allowed—the LED indicator lights in red color then. This PRCD-S+ ensures the following functions:

- residual current protection ($I_{\Delta n} = 30\,\text{mA}$),
- undervoltage protection,
- overvoltage protection,
- switching on the PE conductor is performed before switching on conductors L and N and disconnecting is performed in the reverse order; if dangerous voltage/current from the external circuit occurs on the PE conductor, the system disconnects conductors L and N but the PE conductor circuit remains closed with monitored potential.

3.2.5 Socket-outlet RCD

There are many nonmodernized installations of TN-C type but, unfortunately, this type of installation excludes the use of RCDs. As a temporary solution, the Socket-Outlet Residual Current Device (SRCD) can be applied, which enables the increase in the level of safety of the current-using equipment supplied especially from the TN-C type final circuits (Fig. 3.20).

In the SRCD, the residual current protective function is integrated with the socket-outlet function. The SRCD looks similar to the ordinary socket-outlet and enables the installation in TT, TN-S, as well as in TN-C systems [25]. In a TN-C system the PE and N inlet terminals are connected together. The SRCD has a TEST button and ON/OFF function.

3.2.6 Ground fault circuit interrupter (GFCI) with the neutral-to-earth fault detection

A Ground Fault Circuit Interrupter is basically a kind of residual current device used in North America. However, its construction may significantly differ from those implemented in European RCDs. Fig. 3.21 presents an example construction of the GFCI. It is composed of two current transformers: transformer no. 1 is responsible for the detection of the earth fault (residual) current, whereas transformer no. 2 is designed for transformation of the high-frequency test current I_f in the case of the neutral-to-earth fault. Such a fault interferes with the correct operation of the GFCI. To detect the

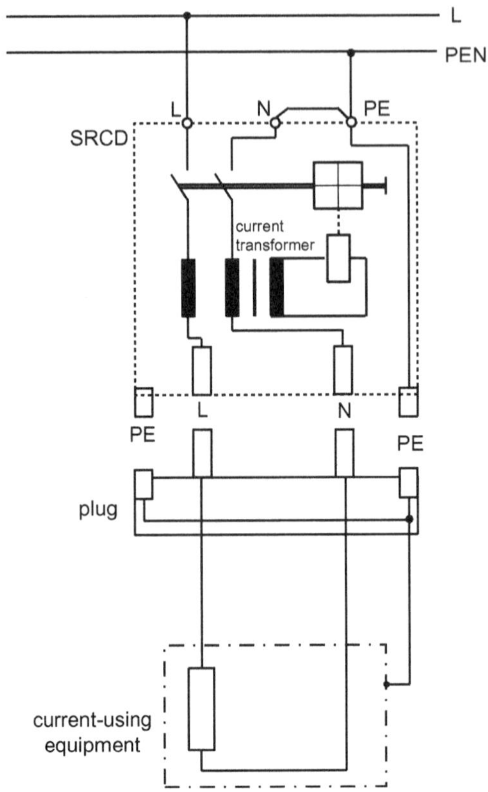

Fig. 3.20 General structure of the socket-outlet RCD.

neutral-to-earth fault, the following solution is implemented. In the secondary circuit of the GFCI, a high-frequency (e.g., 5 kHz) generator (6) is installed, integrated with amplifier (5) and supplied from the main circuit. In the case of the neutral-to-earth fault, the generated current I_f is transformed to the primary side/main circuit via the current transformer (2) and flows in the loop composed of the protective conductor PE, neutral conductor N, and main equipotential bonding busbar MEB, as in Fig. 3.21. This current is detected as a residual current and makes operation of the GFCI. Until there is a short-circuit between the neutral conductor and the earth, the test current does not flow because the aforementioned loop is not closed.

3.2.7 Magneto-optic RCD

In recent years, the nonlinearity of the ferromagnetic iron core of the current transformers has become a problem because high-frequency or

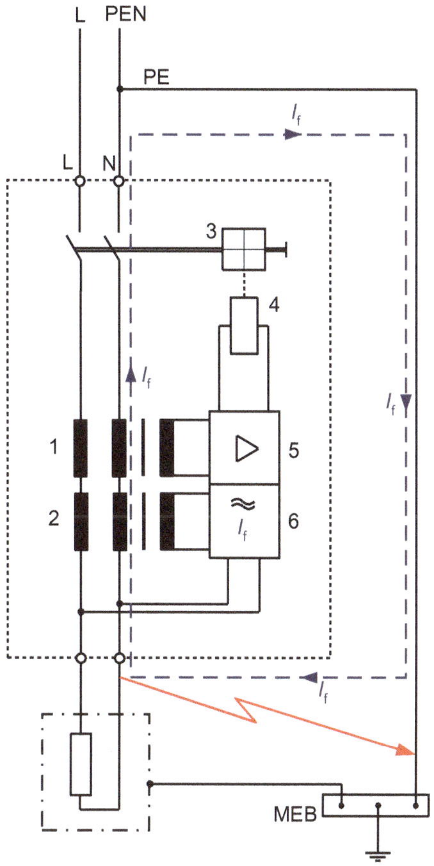

Fig. 3.21 GFCI construction with possibility of the neutral-to-earth fault detection; according to [26]: MEB—main equipotential bonding busbar; I_f—high-frequency test current generated by systems 5, 6; 1—current transformer for detection of the earth fault (residual) current; 2—auxiliary current transformer for generating of the high frequency test current I_f; 3—switching mechanism; 4—relay; 5—amplifier; 6—electronic system for generation of the test current I_f.

nonsinusoidal currents are very often met in modern electrical installations. Such currents may not be transformed properly (see Chapter 8) and an RCD may not trip in the case of the electric shock hazard.

The search for new techniques for residual current detection, ensuring linearity of the residual current converter and other benefits (reduction in size and cost), began to bring promising results from the time when attention was drawn to the possibility of using the linear magneto-optical Faraday effect, for many years used in high-voltage current transformers. In the year

1998, for the detection of low-value residual currents, a new method was used, based on differential fields generated by two coils wound around a Double Extra Dense Flint glass rod (DEDF 785258) [27]. Unfortunately, it was not practical, because the length of the glass rod equal to 10 cm was too long for the application in compact RCDs. Another concept, mentioned in [28], assumed utilization of a CdMnTe diluted magnetic semiconductor crystal—it enabled for reducing the system dimensions but, unfortunately, was expensive.

An interesting solution is presented in the paper [28] as well. This is a Magneto-Optic Residual Current Device (MORCD) based on ferrofluids, which are strongly polarized under the external magnetic field. In this solution, for residual current detection, the electromagnetic circuit generates a residual magnetic field, which in turn induces an observable Faraday effect within the magneto-optical material. In the experiment, a ferrofluid sensor was placed inside the coil (3.5 cm diameter, 1 cm long, 30 turns). The arrangement was located between two linear light polarizers whose transmission axis has an offset equal to 45°. It was illuminated with linear polarized light generated by a 0.5 mW HeNe laser and detected by the photodetector OPT 101. The experimental verification has shown that a very low value of the residual current (1.51 mA) has been properly detected.

3.3 Failures of residual current devices

Reliability of RCD operation has always been one of the fundamental issues analyzed by the maintenance staff and especially by the RCDs' manufacturers. For several decades, in situ tests have been carried out, i.e., at the place of RCD installation, in real conditions of use, in some cases supplemented by laboratory tests of dismounted defective RCDs to recognize the cause of their damage. There are many factors leading to RCD failures: unfavorable ambient conditions (high humidity, dust, low temperature, vibrations), internal defects (secondary winding or relay damage), improper connection of the RCD to the installation. Every defect may lead to the missing operation of the protective device and, hence, to the ineffective protection against electric shock.

One of the indicators of the systems/devices reliability is the failure rate expressed in *Failures In Time* (FIT). The FIT rate indicates the number of failures that can be expected within one billion (10^9) device-hours of operation. For example, if the analyzed population is composed of 10,000 RCDs and 10 RCDs are damaged within one year, the FIT rate is as follows:

$$\lambda = \frac{10\text{RCDs(damaged)}}{10000\text{RCDs(total)} \times 1\text{year}} = \frac{10\text{RCDs(damaged)}}{10000\text{RCDs(total)} \times 8760\text{h}}$$
$$= 1.14 \times 10^{-7}\frac{1}{\text{h}} = 114 \times 10^{-9}\frac{1}{\text{h}} = 114\text{FIT} \quad (3.1)$$

Analyses of the RCD failure rate have been performed in Austria, Germany, and Italy in the period 1989–91 (Tables 3.6–3.9). Results of the verifications of the electrical installations (mainly household) in Austria are included in Table 3.6. There were 3231 verified RCDs installed over the years 1971–90. From this total amount, 207 RCDs had defects, which give an average percentage of the damaged RCDs equal to 6.4%. The failure rate in FIT ranges from 564 to 1141—it qualifies the tested population as highly susceptible to damage.

Similar verifications performed in Germany have shown (Table 3.7) that the RCDs installed there have better reliability indicators. From the total amount of the tested 11,000 RCDs, 274 were damaged—it gives the percentage equal only to 2.5%. The failure rate in FIT ranges from 242 to 437, which is over two times lower than for those installed in Austria.

Verifications performed in Italy have covered around 21,000 RCDs installed up to the year 1991. From this amount, 3866 were installed in

Table 3.6 Failure in time rate (FIT) and percentage of the damaged RCDs verified in Austria within the period 1989–91. Total tested RCDs installed within the period 1971–90: 3231; according to [29].

Years of the RCD production and installation							
1971–75		**1976–80**		**1981–85**		**1986–90**	
FIT	%	FIT	%	FIT	%	FIT	%
564	8.9	816	9.3	586	4.1	1141	3.0

Table 3.7 Failure in time rate (FIT) and percentage of the damaged RCDs verified in Germany within the period 1989–91. Total tested RCDs installed within the period 1971–90: 11000; according to [29].

Years of the RCD production and installation							
1971–75		**1976–80**		**1981–85**		**1986–90**	
FIT	%	FIT	%	FIT	%	FIT	%
437	6.9	404	4.6	242	1.7	418	1.1

Table 3.8 Failure in time rate (FIT) and percentage of the damaged RCDs verified in Italy within the period 1989–91. Total tested RCDs installed up to the year 1991 in household installations: 3866; according to [29].

Years of the RCD production and installation							
-1975		**1976–80**		**1981–85**		**1986–90**	
FIT	%	FIT	%	FIT	%	FIT	%
–	6.6	864	10.6	1154	9.1	1174	3.6

Table 3.9 Failure in time rate (FIT) and percentage of the damaged RCDs verified in Italy within the period 1989–91. Total tested RCDs installed up to the year 1991 in industrial and agricultural installations as well as service workshop installations: 15869; according to [29].

Years of the RCD production and installation							
-1975		**1976–80**		**1981–85**		**1986–90**	
FIT	%	FIT	%	FIT	%	FIT	%
–	14.6	1223	15.0	1306	10.3	1533	4.7

household installations. It is worth mentioning that the total amount of RCDs installed in Italian installations at that time was around 45×10^6. There were voltage-independent verified RCDs installed in various ambient conditions and in various periods before the year 1991. Table 3.8 presents the results of the calculation of the failure in time rate and percentage of the damaged RCDs installed in household circuits. From the 3866 tested RCDs, 274 had defects (7.1%).

The effect of the unfavorable ambient conditions is visible after a comparison of the results from Table 3.8 vs. Table 3.9. RCDs installed in industrial and agricultural installations as well as service workshop installations have relatively higher values in both the FIT indicator and percentage of the damaged RCDs. In the group of the 15,869 tested RCDs, for 1184 of them (7.5%) failures were noted.

Detailed analysis of the verified Italian installations has shown that in outdoor circuits other than the household type, RCDs with defects accounted for 11% of devices. Moreover, in the installations exposed to water splashes, the percentage of defected RCDs was as high as 25%. In wet household installations, 8.7% of RCDs were with defects. On the other hand, in the interiors (both household type and others) in which RCDs were not exposed to high humidity, insolation, vibrations, or corrosive atmosphere, defects were noted for only 6% of RCDs. The verifications have also shown that in 24% of the tested RCDs, the pushing of the TEST button gave a negative result. Therefore it proves that regular use of the test button is necessary—it allows for early detection of the RCD malfunctions.

In addition to the studies described before, many others referring to RCDs have been conducted in the world. Selected results are presented as follows:

Algeria

– *Years: 1985–86*

The effectiveness of protection against electric shock was verified in 80 electrical installations equipped with the main RCD [29]. These installations included 50 RCDs of $I_{\Delta n}=30\,\text{mA}$, 29 RCDs of $I_{\Delta n}=500\,\text{mA}$, and 1 RCD of $I_{\Delta n}=1000\,\text{mA}$. From this amount, 3 RCDs (3.75%) were with detects but in the significant number of the circuits the effectiveness of protection against electric shock was not ensured, due to a lack of the protective conductor or wrong connection in the earthing system.

Austria

– *Year: 1977*

In the year 1977, agriculture- and industrial-type installations were verified [29]. Around 60% of them were utilized for over 10 years. These installations contained 376 RCDs. In 24 RCDs defects were found out—it is 6.4%.

– *Years: 1979–88*

Six thousand two hundred and eighty-eight RCDs were tested in the aforementioned period [29]. 4.4% were with defects of the following nature:

- no tripping under the TEST button pushing (1.4%)—it was recognized mainly in RCD of older constructions,
- the real tripping current higher than the upper permissible value (in those years $1.2I_{\Delta n}$); noted for 0.1% tested RCDs,
- not tripping even under very high value of the residual current (2.9%).

– *Years: 1989–90*

There were 1232 RCDs (986 in household installations, 185 in agricultural premises, 61 in services) tested, which operated from 1 year till 34 years [29]. Protection against electric shock was ineffective in 5.8% of the cases. In 1.3% of the cases, the protection against electric shock was effective but the TEST circuit operated improperly. Results of the verifications are included in Table 3.10.

Table 3.10 Results of RCD verification in Austria within the period 1989–90; according to [29].

		Year		Total	
Description		**1989**	**1990**	**Amount**	**%**
Total tested RCDs		595	637	1232	100%
RCD correct		547	598	1145	92.9%
RCD with defect	TEST circuit with defect	12	4	16	1.3%
	Ineffective protection against electric shock	36	35	71	5.8%
	Tripping circuit of the RCD with defect		21		
Earthing system with defect			14		

Denmark

– *Year 2009*

The report [12] delivers information about 996 tested RCDs:

- 560 of the AC-type and $I_{\Delta n} = 30\,\text{mA}$,
- 400 of the A-type and $I_{\Delta n} = 30\,\text{mA}$,
- 6 of the AC-type and $I_{\Delta n} = 500\,\text{mA}$.

The test indicated that 7% of the RCDs had defects in general. From the group of 560 RCDs of the AC-type and $I_{\Delta n} = 30\,\text{mA}$, 61 RCDs (10.9%) did not operate properly. For the A-type and $I_{\Delta n} = 30\,\text{mA}$ it was 2.0%, accordingly (8 RCDs from 400). Comparing these results (AC-type vs. A-type) it has to be commented that A-type RCDs were newer than the AC-type—from the year 1994, the application of the AC-type RCDs in Denmark is excluded. Tables 3.11 and 3.12 present detailed results of the test. The results prove that hard ambient conditions (here: cold or dusty atmosphere) influence the RCD operation negatively.

France

– *Years: 1970–71*

In the years 1970–71, there were 17,400 RCDs tested of the following rated residual operating currents: 30 mA and 16 mA [29]. They were installed 2 or 3 years before. During the test, it was fund out that 466 RCDs (3.7%) were with defects. In this group:

- 182 RCDs had mechanical defects,
- 449 RCDs had the real tripping current higher than the rated one,
- 15 RCDs had other defects.

Table 3.11 Results of RCD verification in Denmark—the number of RCDs with defect according to the place of their installation; according to [12].

Location	Total tested RCDs Amount	RCDs with defects	
		Amount	%
Household	866	61	7.0
AC-type RCD	511	55	10.8
A-type RCD	355	6	1.7
Industry	28	2	7.7
AC-type RCD	12	1	8.3
A-type RCD	16	1	6.3
Trade, craft, service	32	3	9.4
AC-type RCD	18	2	11.1
A-type RCD	14	1	7.1
Holiday houses and similar	3	0	0
AC-type RCD	1	–	–
A-type RCD	2	–	–
Other	31	3	9.7
AC-type RCD	18	1	5.6
A-type RCD	13	2	15.4

Table 3.12 Results of RCD verification in Denmark—the number of RCDs with defect according to the ambient conditions; according to [12].

Ambient conditions	Total tested RCDs Amount	RCDs with defects	
		Amount	%
Dry	867	61	7.0
AC-type RCD	510	56	10.9
A-type RCD	357	5	1.4
Wet	57	3	5.3
AC-type RCD	27	1	3.7
A-type RCD	30	2	6.7
Cold	6	2	33
AC-type RCD	4	1	25
A-type RCD	2	1	50
Cold and dusty	11	2	18
AC-type RCD	6	2	33
A-type RCD	5	0	0
Dry and dusty	7	1	14
AC-type RCD	5	1	20
A-type RCD	2	0	0

In the group of the RCDs of $I_{\Delta n} = 30$ mA, having real tripping current $I_{\Delta\text{-real}}$ higher than the rated residual operating current $I_{\Delta n}$, the following results had been obtained:

- $I_{\Delta\text{-real}} \leq 35$ mA for 60% of the tested RCDs,
- $35 \text{ mA} < I_{\Delta\text{-real}} \leq 50$ mA for 25% of the tested RCDs,
- $I_{\Delta\text{-real}} \geq 50$ mA for 15% of the tested RCDs.

In the group of the RCDs having $I_{\Delta\text{-real}} \geq 50$ mA, over 50% tripped correctly after manual opening and reclosing the RCD. This indicates the mechanical causes of incorrect tripping. Therefore it underlines a very important issue, which is periodically pushing the TEST button.

Germany

- *Years: 1968–71*

  In Germany, in the year 1966, VDE regulations introduced obligatory use of RCDs in construction and demolition sites [29]. Two years later verification of 281 such installations in Bavaria was performed. Among 7499 circuits with RCDs, as many as 1878 did not ensure protection against electric shock. At this point, it has to be mentioned that most of defects referred to the lack of continuity of the protective conductor or failures in earthing systems. It was the high amount of installations with RCDs in which the protection against electric shock was ineffective. Therefore, after the verifications, detected defects were quickly removed. In the consecutive years, the percentage value of defected circuits with RCDs decreased, as presented in Table 3.13.

- *Years: 1976–77*

  In the aforementioned period, 7292 agricultural installations were verified [29]. These verifications were composed of the RCD testing, the earthing systems testing as well as wiring testing. The verified installations and RCDs were adopted as damaged if permissible touch voltage was exceeded (due to defects in RCDs, broken protective conductor, too high value of the resistance to earth). The results of the verifications are presented in Tables 3.14, 3.15 and 3.16. As for the number of RCDs with defects (Table 3.15), 3.37% of them were adopted as not suitable for further utilization.

- *Years: 1985–96*

  During such a long period, almost 300,000 RCDs were verified in agricultural premises [29]. Every year delivered results for around 24,000–25,000 devices. The following defects were registered:

  - bypassed RCDs,
  - TEST circuit operated improperly,

Table 3.13 Results of the verifications of the construction and demolition sites installations with RCDs in Germany within the period 1968–71; according to [29].

The year of the verification	1968	1969	1970	1971
Number of the verified circuits with RCDs	7499	6366	8544	8139
Number of the circuits with defects	1878	1274	1237	981
Defects in (%)	25	20	14.5	12

Table 3.14 Results of the verifications of the agricultural installations (7292) in Germany within the period 1976–77; according to [29].

Description		Amount	%
RCD correct, effective protection against electric shock		6685	91.85
Ineffective protection against electric shock	RCD with defect	246	3.37
	Defect in installation (RCD correct)	418	5.73

Table 3.15 Defects in RCDs in agricultural installations (7292) in Germany within the period 1976–77; according to [29].

Description	Amount	%
Real tripping current of the RCD higher than $I_{\Delta n}$	91	1.25
No tripping of the RCD under TEST button pushing	155	2.13
Total	246	3.37

- other defects (mainly: tripping circuit operated improperly).

If at least one of the above-mentioned defects occurred, the RCD was qualified as damaged. Results of the verification are presented in Table 3.17.

– *Years: 1988–91*

Verifications performed in the period 1988–91 referred to 43,000 RCDs installed in various types of buildings: industry, agricultural, services [29]. RCDs were tested in the place of the installation as well as in laboratory conditions. 1100 devices (2.6%) had defects.

– *Year: 2000*

There were tested voltage-independent RCDs installed in Bavarian agricultural premises [29]. Data referring to 150,000 tested RCDs showed that around 5% of them had defects.

Table 3.16 Defects in agricultural installations (7292) in Germany within the period 1976–77; RCDs were correct; according to [29].

Description	Amount	%
Too high value of the resistance to earth	219	3.00
Short-circuit of the conductors N and PE	169	2.32
Other sources of the ineffective protection against electric shock	30	0.41
Total	418	5.73

Table 3.17 Results of RCD verification in agricultural premises in Germany within the period 1985–96; according to [29].

Year	Tested RCDs Amount	RCDs with defects	
		Amount	%
1985	26,361	742	2.81
1986	25,736	866	3.36
1987	22,433	757	3.37
1988	23,327	716	3.07
1989	23,963	585	2.44
1990	23,843	741	3.10
1991	24,234	822	3.39
1992	24,799	881	3.55
1993	24,451	964	3.94
1994	24,484	1145	4.67
1995	23,819	943	3.96
1996	23,819	990	4.15
Total	291,335	10,152	3.5

Holland

- *Year: 2000*

  In Holland, 634 voltage-dependent RCDs with integral overcurrent protection were tested at the place of their installation [29]. The RCDs were from one manufacturer. The result was very optimistic—only 1 RCD had a defect. It is merely 0.16% of the verified population.

Poland

- *Years: 1970–71*

  In Poland, RCDs on ships under construction, in temporary installations were tested [29]. Such installations give very hard ambient conditions for RCDs. The protective devices installed in portable distribution boards were exposed to vibrations, low temperature, and high humidity. Three groups of voltage-independent RCDs of a rated residual operating current $I_{\Delta n} = 30\,mA$ were tested:

(1) 125 RCDs utilized 3 years,

(2) 111 RCDs utilized 1.5 years,

(3) 86 RCDs utilized 5 years.

In the group of the 125 RCDs utilized 3 years, over 50% (72 RCDs) were required to be repaired at least twice. Only 13 RCDs had not got any problems. The evaluated failure rate for all analyzed population was as much as 12,000 FIT! It was over 10 times higher than the average value for RCDs installed in those years.

– *Year: 2017*

In the year 2017, Electrical Apparatus Producers Section of the Polish Chamber of Commerce for Electronics and Telecommunications (SPAE KIGEiT) in cooperation with Quality Research Office (BBJ), operating in the area of conformity assessment of electrical products, tested 23 new RCDs from various manufacturers [30]. The aim of the test was to check the RCDs conformity with standard IEC/EN 61008-1 [2]. The test showed that 13 RCDs (57%) did not meet the requirements of this standard. Requirements of the following clauses of the IEC/EN 61008-1 were not fulfilled:

- 9.7.3 Dielectric strength of the main circuit—1 RCD did not meet the requirement,
- 9.7.7 Verification of impulse withstand voltages (across clearances and across solid insulation) and of leakage current across open contacts—4 RCDs did not meet the requirement,
- 9.11.2.2 Verification of the rated making and breaking capacity (I_m)—11 RCDs did not meet the requirement,
- 9.19.2 Verification of behavior at surge currents up to 3000 A (8/20 μs surge current test)—1 RCD did not meet the requirement.

The aforementioned test has shown that the quality and reliability of even new products can be doubtful.

South Africa

– *Year: 1996*

Results of the test presented in [31] show that from among 17,000 tested RCDs, the percentage of the RCDs with defects varies within the range from 2.5% to 7.5% and it is similar to those noted in Europe. However, if we compare the voltage-independent RCDs with the voltage-dependent RCDs having electronics components, the latter are more reliable. Voltage-independent RCDs with electromechanical relay have often problems with their magnetic armature, in the case of the dusty atmosphere or high humidity.

USA

– *Year: 2001*

In the USA as an equivalent to RCD, a GFCI is used. The test referred to 2860 GFCIs installed within the years 1974–98 in household locations [29]. The verified GFCIs were a socket-outlet type or installed in the distribution board. Both types were voltage dependent with electronic equipment. Verification covered 12 cities located in various climatic conditions. Interestingly, in dry climatic conditions the percentage of the GFCIs with defect was lower:

- Dry condition: 7.5% (7.3%—socket-outlet type, 11%—distribution board type),
- Wet condition: 11.1% (10.5%—socket-outlet type, 17.5%—distribution board type).

Most of the defects concerned electronic elements of the GFCI.

References

[1] IEC 60755:2017, General safety requirements for residual current operated protective devices, 2017.
[2] IEC 61008-1:2010, Residual current operated circuit-breakers without integral overcurrent protection for household and similar uses (RCCBs)—Part 1: General rules, 2010.
[3] IEC 61009-1:2010, Residual current operated circuit-breakers with integral overcurrent protection for household and similar uses (RCBOs)—Part 1: General rules, 2010.
[4] S. Czapp, The effect of earth fault current harmonics on tripping of residual current devices, in: International School on Nonsinusoidal Currents and Compensation, IX Conference-Seminar (ISNCC) Łagów, Poland, 10–13 June, 2008.
[5] S. Czapp, The impact of DC earth fault current shape on tripping of residual current devices, Elektronika ir Elektrotechnika 4 (84) (2008) 9–12.
[6] S. Czapp, The effect of PWM frequency on the effectiveness of protection against electric shock using residual current devices, Przeglad Elektrotechniczny 1 (87) (2011) 24–27.
[7] Czapp S., Guzinski J.: Electric shock hazard in circuits with variable-speed drives, Bull. Pol. Ac.: Tech., 2018, no. 3 (66), pp. 361–372.
[8] S. Czapp, K. Dobrzyński, J. Klucznik, Z. Lubośny, R. Kowalak, Improving sensitivity of residual current transformers to high frequency earth fault currents, Archives of Electrical Engineering 66 (3) (2017) 485–494.
[9] S. Czapp, K. Dobrzyński, J. Klucznik, Z. Lubośny, Low-frequency tripping characteristics of residual current devices, in: 2017 IEEE International Conference on Environmental and Electrical Engineering & 2017 IEEE Industrial and Commercial Power Systems Europe (EEEIC /I&CPS Europe), 6–9 June, 2017, Milan, Italy, 2017, pp. 298–301.
[10] S. Czapp, S. Szultka, Tripping limitations of residual current devices in photovoltaic installations, in: 18th International Scientific Conference on Electric Power

Engineering (EPE), 2017, Kouty nad Desnou, Czech Republic, 17–19 May, 2017, 2017, pp. 1–5.
[11] IEC 62423:2009, Type F and type B residual current operated circuit-breakers with and without integral overcurrent protection for household and similar uses, 2009.
[12] Esbjerg, Field test of installed residual current devices in Denmark, Esbjerg, Denmark, 14, 28 Oct. 2009.
[13] DFS 4 016-2/0,03-B+, Data Sheet, Doepke, 2014.
[14] Allstromsensitive Fehlerstrom-Schutzschalter F200 B/B + und F800 B/B + Produktinformation, ABB STOTZ-KONTAKT GmbH, 2011.
[15] IEC TS 60479-1:2016 Effects of current on human beings and livestock. Part 1: General aspects.
[16] IEC TS 60479-2:2007 Effects of current on human beings and livestock. Part 2: Special aspects.
[17] ABB, The emergency stop according to ABB. Technical, ABB, Day by DIN 1|14, pp. 68–60.
[18] 230VAC, Remote Switching Unit, Type FSA, Schrack Technik.
[19] Fehlerstromschutz mit Wiedereinschaltung, Elektropraktiker, Berlin 59 (11) (2005) 918.
[20] Gewiss, n.d. https://www.gewiss.com/ww/en/products/experience-catalogue/catalogs/series/power/90-restart-range-automatic-reclosing-devices (access: 2021.05.24).
[21] IEC 63024:2017, Requirements for automatic reclosing devices (ARDs) for circuit-breakers, RCBOs and RCCBs for household and similar uses, 2017.
[22] HD 639 S1:2002, Electrical accessories—Portable residual current devices without integral overcurrent protection for household and similar use (PRCDs), 2002.
[23] H.-J. Fleckenstein, Portable Residual Current Devices: Typen, Schutzumfang Und Praktischer Einsatz, Heinrich Kopp GmbH, 2006.
[24] PRCD-S, PRCD-S+ Mobile personal protection on construction sites, PCE Electric GesmbH. Austria, 2020.
[25] GW 95 921, Flush mounting RCD safety socket-outlet, GW 95 921. 90 RCD, Gewiss, 2015.
[26] G. Biegelmeier, G. Kiefer, K.-H. Krefter, Schutz in elektrischen Anlagen, VDE-Verlag, Berlin-Offenbach, 1999.
[27] L.M. Maniner, P. Adl, R.T. Rakowski, A. Cheshmehdaast, Optical current transducer for residual current sensing in the milliampere range, Sensors Actuators A Phys. 67 (1998) 102–108.
[28] L. Martinez, R. Rakowski, F. Cecelja, Design of a magneto-optic residual current device using aqueous ferrofluid as the sensing material, in: 21st IEEE Instrumentation and Measurement Technology Conference, Como, Italy, 18–20 May, 2004.
[29] Musial E., Czapp S.: Residual current devices—reliability. INPE: Informacje o Normach i Przepisach Elektrycznych, 2008, no. 110–111, pp. 3–40 (in Polish).
[30] Kigeit, 2017, https://kigeit.org.pl/2017/10/17/spae-kigeit-poleca-artykul-nt-badan-wylacznikow-roznicowopradowych-rccb/ (access: 2021.05.24).
[31] V. Cohen, Why electronic and not electromechanical ELCBs? in: Proceedings of IEEE. AFRICON '96, 24–27 Sept., vol. 2, 1996, pp. 715–719.

CHAPTER 4

Protection against electric shock

4.1 Protection in case of fault

4.1.1 General information

Every electrical installation should be designed to fulfill the fundamental safety rule included in standard IEC 61140 [1]—hazardous live parts of electrical installation/equipment shall not be accessible and accessible conductive parts shall not be hazardous live.

The latter provision (accessible conductive parts shall not be hazardous live) is usually fulfilled by application of one of the following measures in case of a fault:

- automatic disconnection of supply,
- double or reinforced insulation,
- electrical separation for the supply of an item of current-using equipment.

Residual current devices are applied in the protective measure automatic disconnection of supply. When insulation fault occurs, disconnection of supply should be ensured within the time specified in standard HD 60364-4-41 [2] (Table 4.1), and the role of the RCD is to switch off this supply automatically.

Disconnection times specified in Table 4.1 refer to the following circuits:

- socket-outlet circuits having a rated current not exceeding 63 A,
- final circuits supplying fixed connected current-using equipment having a rated current not exceeding 32 A.

For other final circuits and for distribution circuits, the following disconnection times may be adopted:

- 5 s in TN systems,
- 1 s in TT systems.

When an RCD is used as a disconnecting device, in TN or TT systems of $120\,V < V_o \leq 230\,V$ the current I_a causing disconnection of supply within the specified time should be taken as in Table 4.2.

Conditions for the effectiveness of the protection by automatic disconnection of supply and proper operation of RCDs depend on the type of low-voltage system—TN, TT, or IT.

Residual Current Devices
https://doi.org/10.1016/B978-0-323-89783-9.00010-2

Table 4.1 Maximum permissible disconnection times (in seconds); according to HD 60364-4-41 [2].

Low-voltage system	$50\,V < V_o \leq 120\,V$		$120\,V < V_o \leq 230\,V$		$230\,V < V_o \leq 400\,V$		$V_o > 400\,V$	
	AC	DC	AC	DC	AC	DC	AC	DC
TN	0.8	[a]	0.4	1	0.2	0.4	0.1	0.1
TT	0.3	[a]	0.2	0.4	0.07	0.2	0.04	0.1

[a]Disconnection may be required for reasons other than protection against electric shock.

V_o—nominal line-to-earth voltage.

Table 4.2 Tripping current of RCDs as a function of the maximum permissible disconnection time; according to [3].

Maximum permissible disconnection times (s)	Current I_a causing disconnection of supply of RCDs within the specified time					
	No-delayed and short-time delayed			Delayed (S-type)		
	AC	A (30 mA)	B	AC	A	B
0.2	$2I_{\Delta n}$	$4I_{\Delta n}$	$4I_{\Delta n}$	$2I_{\Delta n}$	$2.8I_{\Delta n}$	$4I_{\Delta n}$
0.4	$I_{\Delta n}$	$2I_{\Delta n}$	$2I_{\Delta n}$	$2I_{\Delta n}$	$2.8I_{\Delta n}$	$4I_{\Delta n}$
1	$I_{\Delta n}$	$2I_{\Delta n}$	$2I_{\Delta n}$	$I_{\Delta n}$	$1.4I_{\Delta n}$	$2I_{\Delta n}$
5	$I_{\Delta n}$	$2I_{\Delta n}$	$2I_{\Delta n}$	$I_{\Delta n}$	$1.4I_{\Delta n}$	$2I_{\Delta n}$

4.1.2 Residual current devices in TN system

The application of RCDs is allowed only in TN-S system (or TN-S part of the TN-C-S system—Fig. 4.1A). In the case of application of RCDs in TN-C system, no detection of earth fault (residual) current is possible (Fig. 4.1B). Earth fault current $I_{L\text{-}E}$ flows across L1 and PEN conductors, which go through the current transformer of the RCD, and no residual current occurs $I_\Delta = 0$.

Earth fault in a TN system gives a relatively high value of the earth fault current because the earth current path is metallic. This phenomenon is favorable for the detection of current by overcurrent protective devices and even more by RCDs. In TN systems the tripping characteristic of the protective device and the considered circuit loop impedance shall fulfill the following requirement:

$$Z_s \cdot I_a \leq V_o \tag{4.1}$$

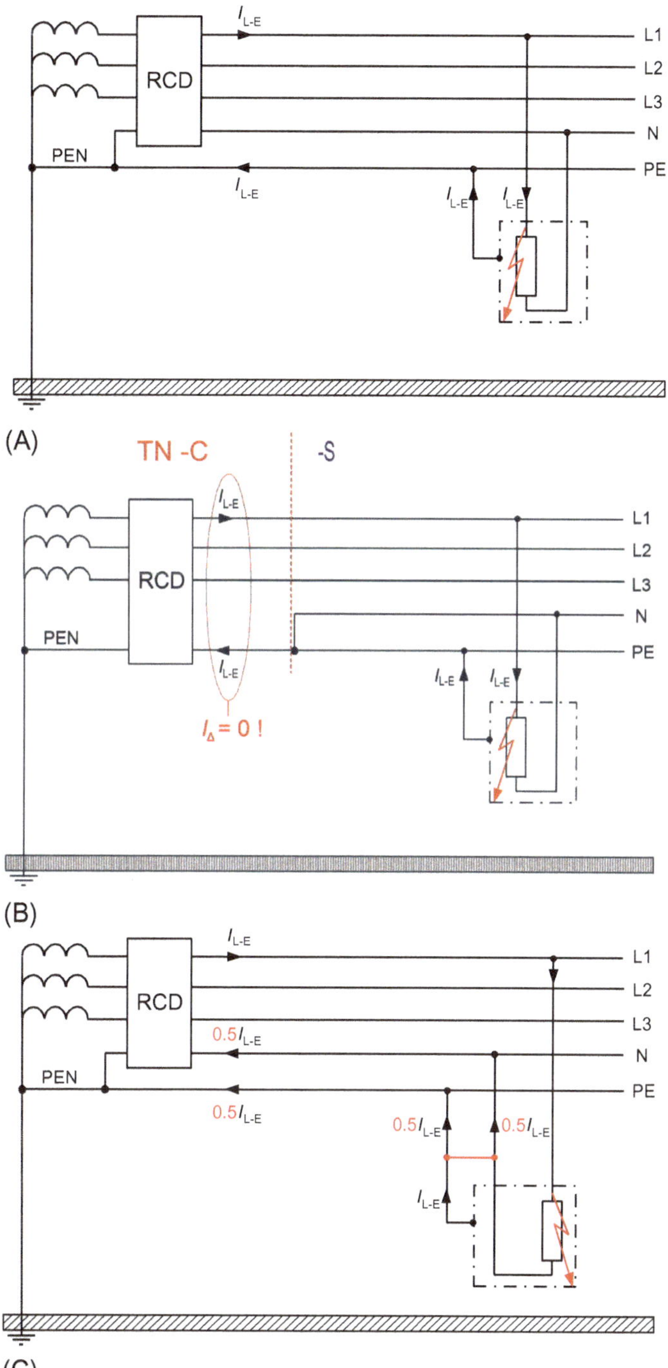

Fig. 4.1 Application of the RCD in a TN-C-S system: (A) in TN-S part of the system (properly), (B) in TN-C part of the system (not allowed), (C) in TN-S part of the system with unintentional connection PE-N downstream the RCD; $I_{L\text{-}E}$—earth fault current, I_{Δ}—residual current.

where

Z_s is earth fault loop impedance composed of the source, the line conductor to point of the earth fault and return path to the source via protective conductor, Ω,

I_a is the current causing disconnection of supply within the specified time, A,

V_o is nominal line-to-earth voltage, V.

Reliable operation of RCDs depends, among others, on the right connections in a circuit. An unintentional connection of PE conductor with N conductor downstream the RCD (Fig. 4.1C) makes the earth fault current $I_{L\text{-}E}$ to be divided into two paths (per approx. $0.5I_{L\text{-}E}$ in PE and N). In effect, it gives deterioration of the RCD sensitivity. Moreover, during normal operating conditions, when there is no earth fault, the load current is also divided into the same two paths, and unwanted tripping of the RCD occurs, if the value of load current is at least twice the rated residual operating current of the RCD.

4.1.3 Residual current devices in TT system

In the case of the earth fault in a TT system, earth fault current flows in fact through the earth (Fig. 4.2). The value of the current is relatively low because it is limited by the resistance of the earth electrode R_A of the current-using equipment and resistance of the earth electrode connected to the source neutral point. The value of these resistances connected in series is usually from a few to several dozen ohms. So, the earth fault current can be less than the rated current of the overcurrent device (e.g., less than 16 A) protecting the considered circuit. For this reason, automatic disconnection of supply is in practice realized with the use of RCDs. The condition to fulfill safety is given by the following dependency:

$$R_A \cdot I_a \leq 50\,\text{V} \tag{4.2}$$

where

R_A is the sum of the resistances of the earth electrode and the protective conductor for the exposed-conductive-parts, Ω,

I_a is the current causing disconnection of supply within the specified time, A.

This condition is easy to fulfill even by an RCD of low sensitivity, e.g., 500 mA. In such a case the required R_A resistance is (assumed no-delayed AC-type RCD and disconnection time 1 s):

$$R_A \leq \frac{50\,\text{V}}{I_a} = \frac{50\,\text{V}}{0.5\,\text{A}} = 100\ \Omega \tag{4.3}$$

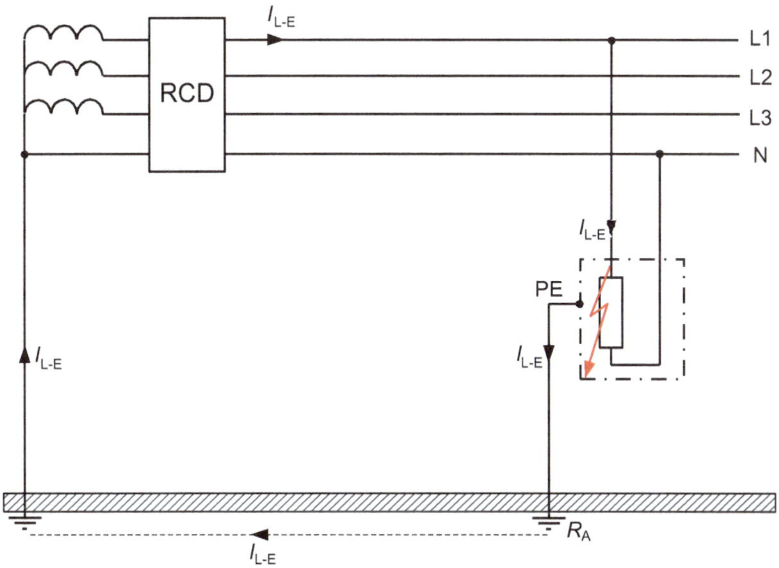

Fig. 4.2 Application of the RCD in TT system; $I_{L\text{-}E}$—earth fault current.

4.1.4 Residual current devices in IT system

In an IT system, when an isolated neutral point is applied, special care is required to be taken during selecting and installing of RCDs. In this system, in the case of the first earth fault, the earth fault current has a very low value as well as capacitive character and depends on the capacitance-to-earth of the network (Fig. 4.3).

Thanks to the low value of the current I_d, automatic disconnection of supply in case of the first fault is usually not required, which gives a high level of continuity of supply. Exposed-conductive-parts shall be earthed individually or in groups, or collectively, and safety condition is given by the following dependency:

$$R_A \cdot I_d \leq 50\,\mathrm{V} \tag{4.4}$$

where

R_A is the sum of the resistances of the earth electrode and the protective conductor for the exposed-conductive-parts, Ω,

I_d is the earth fault current of the first fault, A.

Disconnection of supply in case of the first fault can be required if an IT system is not selected for continuity of supply but due to the necessity of fire/explosion hazard reduction. Special attention has to be taken to the selection of the point of the RCD location.

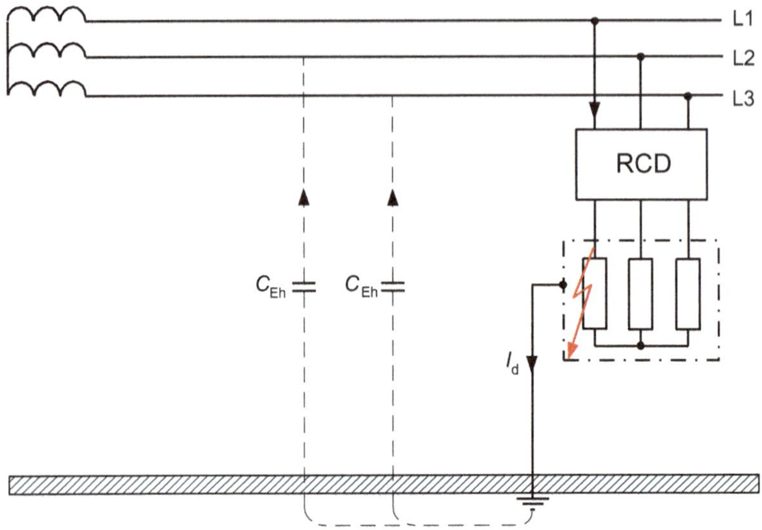

Fig. 4.3 Application of the RCD in IT system; I_d—earth fault current of the first fault, C_{Eh}—capacitance-to-earth of the conductors.

If the RCD is installed at the origin of an IT network (Fig. 4.4), the capacitive current flows through the earth capacitances C_{Eh} in part of the network downstream the RCD. Earth fault current is not detected by the RCD, because this current does not generate the residual current.

No reaction of the RCD installed at the origin of the IT system may occur even in case of double earth fault (high-value current), which may not be accepted in areas with fire/explosion risk due to thermal effect or sparking/arcing effect (Fig. 4.5).

The undesirable reaction of the RCD is expected in the arrangement presented in Fig. 4.6. Earth fault occurs in the part of the installation upstream of the RCD, but earth current is returning mainly downstream the RCD, which generates the residual current and the unwanted tripping.

In some IT systems, earth fault current may be too low to initiate tripping of the RCD in case of the first fault. In such systems, an indirect earthing of the neutral point via a capacitor can be used (Fig. 4.7). Capacitance C_N, giving sufficient value to activate the RCD of $I_{\Delta n}$, can be selected as follows:

$$C_N > \frac{\sqrt{3} \cdot I_{\Delta n}}{\omega \cdot V} \tag{4.5}$$

where

C_N is the capacitance of the neutral point,

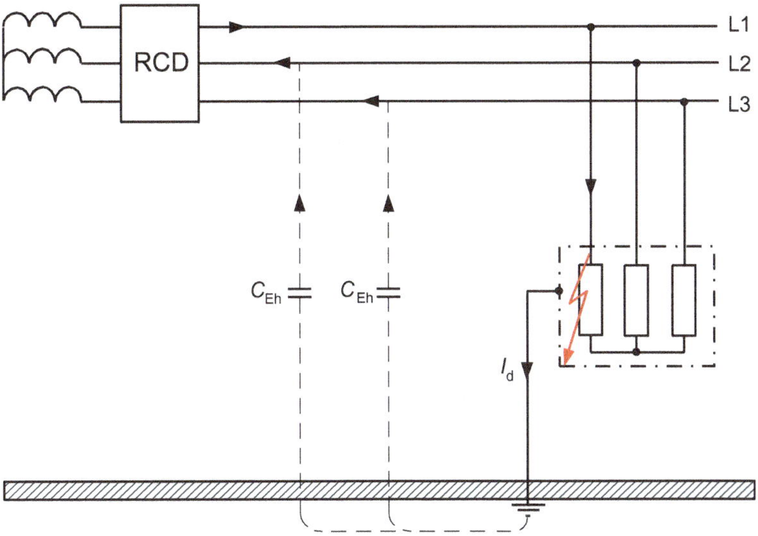

Fig. 4.4 No reaction of the RCD installed at the origin of the IT system; I_d—earth fault current of the first fault, C_{Eh}—capacitance-to-earth of the conductors.

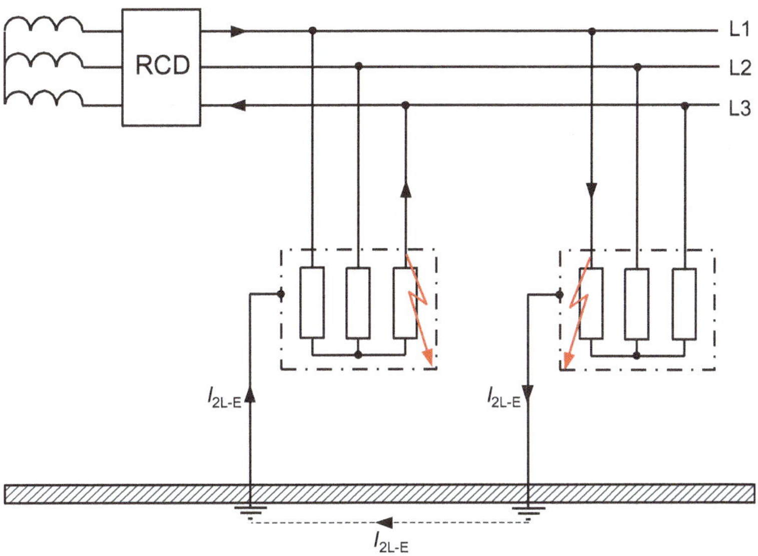

Fig. 4.5 No reaction of the RCD installed at the origin of the IT system, in case of double fault; $I_{2L\text{-}E}$—double earth fault current.

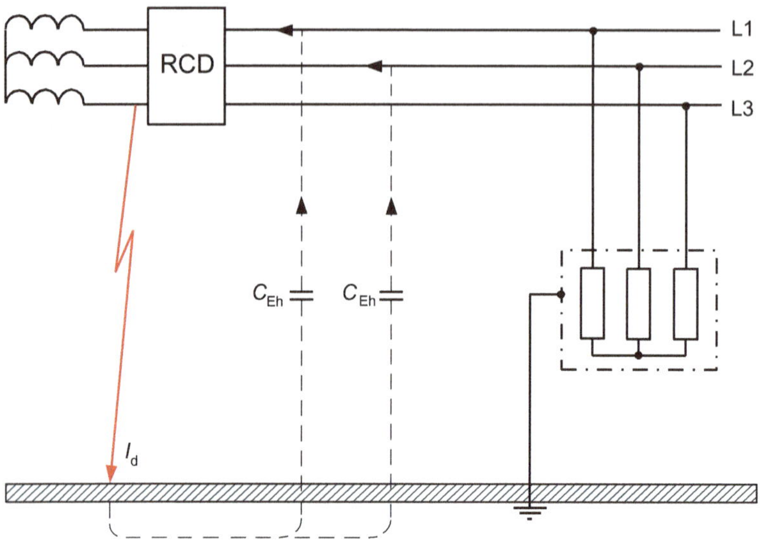

Fig. 4.6 Unwanted tripping of the RCD in an IT system in case of a fault in the part of the circuit not protected by this RCD; I_d—earth fault current of the first fault, C_{Eh}—capacitance-to-earth of the conductor.

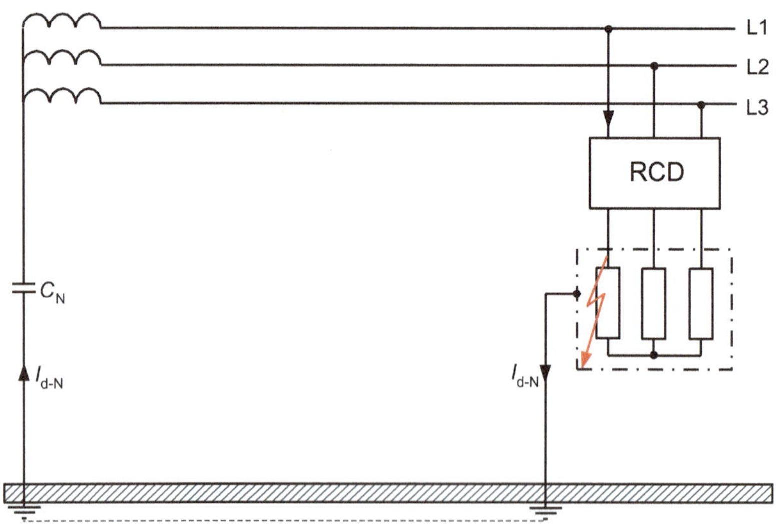

Fig. 4.7 Indirect earthing of the neutral point via capacitor C_N; $I_{d\text{-}N}$—earth fault current of the first fault.

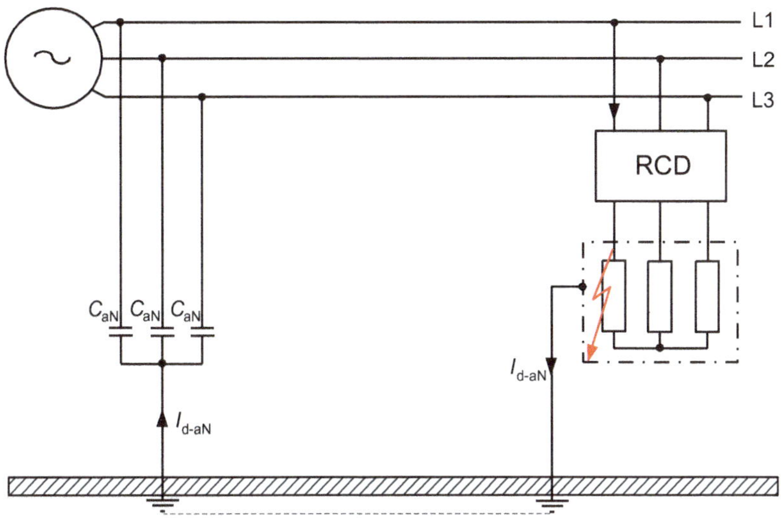

Fig. 4.8 Creating of an artificial neutral point in an IT system; C_{aN}—artificial neutral point capacitance, $I_{d\text{-}aN}$—earth fault current in the network with the artificial neutral point.

$I_{\Delta n}$ is a rated residual operating current of the installed RCD,
V is a nominal voltage of the network,
ω is pulsatance.

If an IT system has no accessible neutral point, improved operation of the RCD may be achieved by creating an artificial neutral point (Fig. 4.8). Capacitance C_{aN} can be selected as follows:

$$C_{aN} > \frac{I_{\Delta n}}{\sqrt{3} \cdot \omega \cdot V} \tag{4.6}$$

where

C_{aN} is the capacitance in the one branch of the artificial neutral point,
$I_{\Delta n}$ is a rated residual operating current of the installed RCD,
V is a nominal voltage of the network,
ω is pulsatance.

4.2 Additional protection

Additional protection in electrical installations is used especially in locations where an increased risk of electric shock occurs (e.g., the use of handheld tools) and/or unfavorable external conditions exists (e.g., wet places). For this purpose, high-sensitivity RCDs are required to be used.

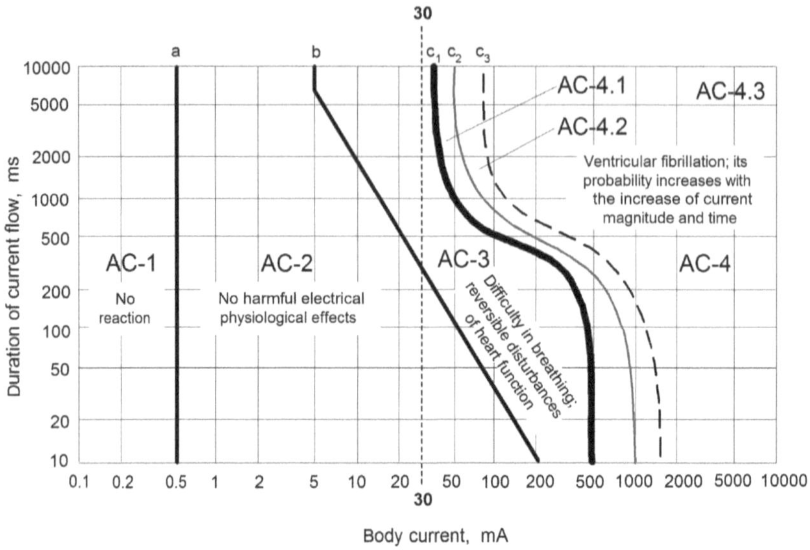

Fig. 4.9 Conventional current-time zones of the effect of AC currents (15–100) Hz on persons; current path: left hand to feet; according to [4].

Present requirements referring to the additional protection with the use of RCDs are based on the fundamental criteria derived from pathophysiological tests. Fig. 4.9 presents the effects of current on persons, according to the report [4].

In zone AC-1 a perception of current is possible but usually, no startled reaction is expected. The perception and involuntary muscular contraction are possible in zone AC-2, but rather no harmful electrical physiological effects occur. In zone AC-3 relatively high likelihood of involuntary muscular contractions and difficulty in breathing exist as well as reversible disturbances of formation and conduction of impulses in the heart and even immobilization are possible. Zone AC-4 may be dangerous for persons—cardiac arrest and breathing arrest may occur, burns or other cellular damage are possible, probability of ventricular fibrillation increases with the increase of current magnitude and time. In the zone AC-4.1 probability of ventricular fibrillation is up to about 5%, in zone AC-4.2—up to about 50%, and in zone AC-4.3—above 50%.

The most important in the chart presented in Fig. 4.9 is trace c_1. For the long-time flow of the body current, the trace slightly exceeds 30 mA. This trace can be called the "threshold of ventricular fibrillation."

For this reason, in many parts of standard HD 60364 "Low-voltage electrical installations" requirements for the application of RCDs of $I_{\Delta n} \leq 30$ mA in selected circuits are included. Such RCDs are considered as additional protection. The use of the RCDs of $I_{\Delta n} \leq 30$ mA is very important especially to prevent persons from the fatal effect of electric shock in case of the carelessness of users when direct contact occurs (contact with a live conductor—Fig. 4.10). There is no insulation fault then, therefore circuit-breakers do not operate due to a relatively low value of current (only body current flows), and only high sensitivity RCDs may detect this dangerous accident.

Provisions of standard HD 60364 referring to the obligation to use RCDs in some circuits, especially as additional protection, are presented in Chapter 6.

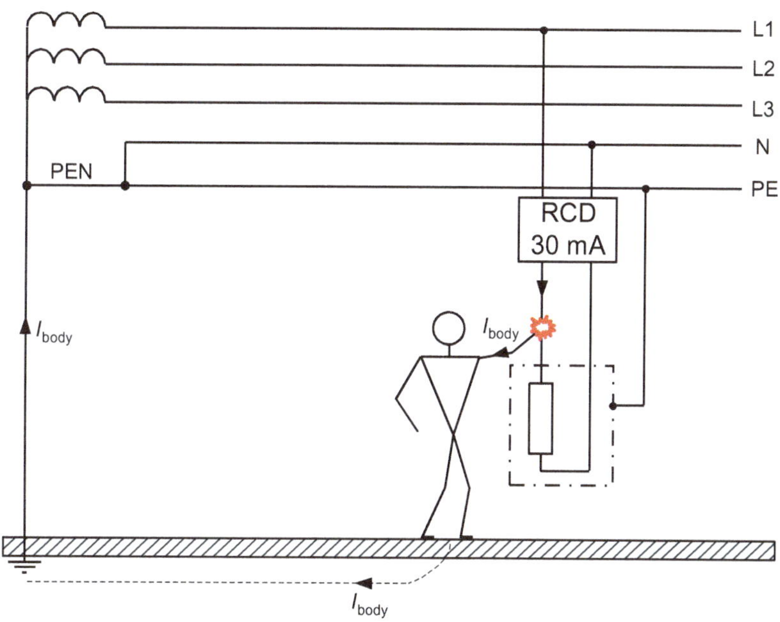

Fig. 4.10 Direct contact in a low-voltage TN system and body current (I_{body}) flow.

References

[1] IEC 61140:2016, Protection against electric shock—Common aspects for installation and equipment, 2016.
[2] HD 60364-4-41:2017, Low-voltage electrical installations—Part 4–41: Protection for safety—Protection against electric shock, 2017.
[3] IEC 61008-1:2010, Residual current operated circuit-breakers without integral overcurrent protection for household and similar uses (RCCBs)—Part 1: General rules, 2010.
[4] IEC TS 60479-1:2016, Effects of current on human beings and livestock. Part 1: General aspects, 2016.

CHAPTER 5

Protection against fire

5.1 Leakage currents

Leakage currents which flow from line conductor to earth, through the resistive medium, generate heat, and in some circumstances, may cause a fire. The amount of heat emitted by this current can be calculated with the use of the following expression:

$$P_{heat} = V_o \cdot I_{leak} \tag{5.1}$$

where

P_{heat} is thermal power generated by current flowing to earth through the resistive medium,

V_o is line-to-earth nominal voltage,

I_{leak} is leakage current.

Such a current can be detected by RCDs, and the risk of fire is eliminated then, provided that their rated residual operating current $I_{\Delta n}$ is relatively low. The permissible value of $I_{\Delta n}$ of RCDs depends on the line-to-earth nominal voltage. Table 5.1 presents maximal values of thermal power which may appear in circuits with RCDs of particular values of the rated residual operating current.

The maximal permissible value of a rated residual operating current for 230/400 V low-voltage networks is given in standard HD 60364-4-42:2011 [1] (Subclause 422.3.9). According to this standard, final circuits and current-using equipment installed in locations, where the risk of fire due to stored materials or technological processes exists, are required to be protected against insulation faults in the following way:

(a) In TN and TT systems, RCDs of $I_{\Delta n} \leq 300\,mA$ shall be installed. In locations where resistive faults may cause a fire (e.g., overhead heating with heating film elements) RCDs of $I_{\Delta n} \leq 30\,mA$ shall be installed.

(b) In IT systems, IMDs (insulation monitoring devices) covering the entire installation, or RCMs (residual current monitoring devices) that monitor the residual current in the final circuits, both with audible and visual signals, shall be used. As an alternative, RCDs may be used—their rated residual operating current should be the same as in point a).

Residual Current Devices
https://doi.org/10.1016/B978-0-323-89783-9.00006-0

Table 5.1 Thermal power produced by leakage current, as a function of nominal line-to-earth voltage and the rated residual operating current of RCDs.

Line-to-earth nominal voltage (V)	Rated residual operating current of RCDs (mA)	Thermal power (W)
120	30	3.6
	100	12
	300	36
	500	60
230	30	7
	100	23
	300	**69**
	500	115
400	30	12
	100	40
	300	120
	500	200

These requirements also apply to the circuits going through these locations.

Requirements for protection against fire are also included in standard HD 60364-5-53:2015 [2] (Clauses 532.2 and 532.3). This standard requires the installation of RCDs of $I_{\Delta n} \leq 300\,\text{mA}$ at the origin of the protected circuits. In IT systems, residual current monitoring devices (RCMs) for preventing the risk of fire may be used (alternative to RCDs), provided that the area is supervised by instructed or skilled persons. These devices also have to be installed at the origin of the protected circuits. RCMs have to provide audible and visual signals, and their rated residual operating current shall not exceed 300 mA.

The aforementioned provisions of the standards indicate that for protection against fire, RCDs having a rated residual operating current not exceeding 300 mA shall be used (bold font in Table 5.1). In such a case, the maximal generated thermal power is equal to 69 W. This requirement is valid for low-voltage installations of nominal line-to-earth voltage equal to 230 V. In installations of other nominal voltages, maximal permissible value of $I_{\Delta n}$ is different. In order not to exceed 69 W (as in a system of $V_o = 230\,\text{V}$), in a system of $V_o = 120\,\text{V}$ a rated residual operating current of RCDs may have even 500 mA (it gives max thermal power 60 W—see Table 5.1), but in a system of $V_o = 400\,\text{V}$ this current is limited to 100 mA. In the 400 V system an RCD of $I_{\Delta n} \leq 300\,\text{mA}$ would generate an unacceptable value of power equal to 120 W (400 V × 0.3 A).

Residual current devices, however, may be ineffective in some types of insulation faults, in spite of the fact that real fire hazard exists. The example cases of such a risk are presented in Fig. 5.1 as well as in Section 5.2.

When the insulation of conductors or insulation of current-using equipment is being progressively deteriorated in all three phases of a three-phase circuit, it may lead to the simultaneously flow of leakage currents from all phases (Fig. 5.1).

The resultant value of current which is seen by an RCD depends on the value of particular leakage current flowing from each phase. In the case of unsymmetrical leakage currents (Fig. 5.2A), the resultant value of

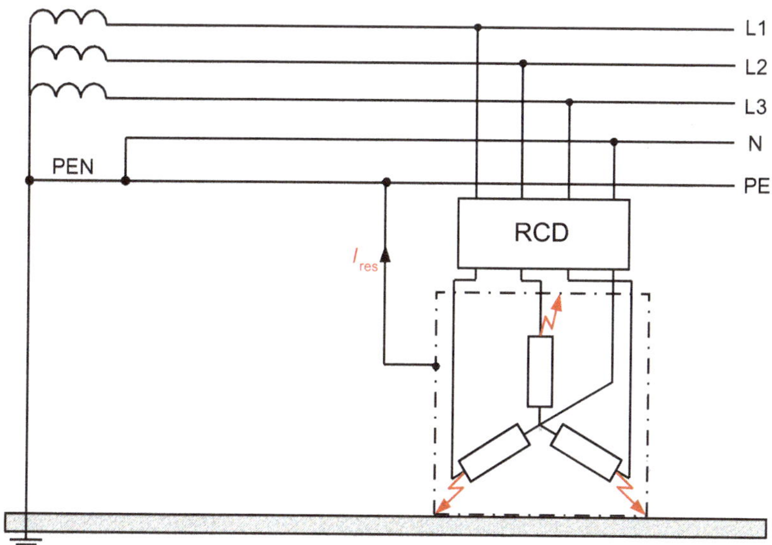

Fig. 5.1 Low-current earth faults in three phases which may not be detected by an RCD; I_{res}—resultant (residual) current.

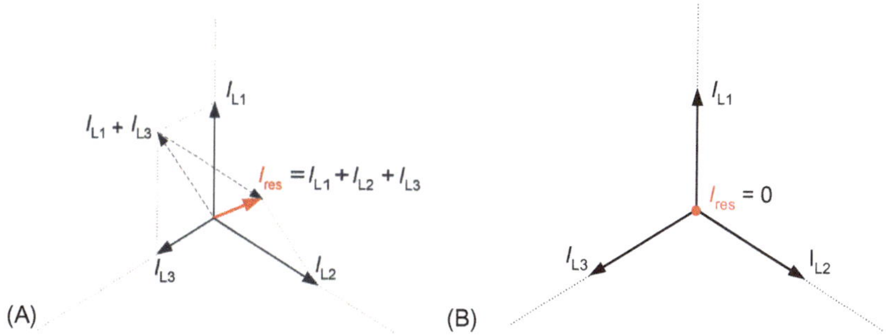

Fig. 5.2 Resultant current I_{res} in case of: (A) an unsymmetrical earth fault in three phases, (B) a symmetrical earth fault in three phases ($I_{res} = 0$); I_{L1}, I_{L2}, I_{L3}—leakage currents in phases L1, L2, and L3, respectively.

current I_{res} is not equal to zero but can be too low to be detected by the RCD. However, the worst case is presented in Fig. 5.2B—insulation-to-earth in each phase can be in very poor condition, leakage currents I_{L1}, I_{L2}, I_{L3} have relatively high value, but the RCD is not able to detect such a fault.

5.2 Arcing

The other phenomenon generating fire risk in electrical installations is arcing. Arcing may occur between live conductors, between live conductor and earth, or may be generated only in one conductor. From the point of view of protection against fire in low-voltage electrical installations, standard [3] recognizes the following types of arc faults:

- parallel arc fault—arc fault where the current is flowing between active (live) conductors in parallel with the load of the circuit,
- earth arc fault—arc fault where the current is flowing from active (live) conductor to earth,
- series arc fault—arc fault where the current is flowing through the load of the final circuit.

Fig. 5.3 presents the aforementioned types of arcing.

If a fault with the electric arc occurs between the line conductor L1 and the neutral conductor N (Fig. 5.3A), the miniature circuit-breaker (MCB) may operate provided that the value of the fault current is high enough to activate an overcurrent tripping mechanism (thermal or electromagnetic) of the MCB. In the case of a short-circuit between the line conductor L1 and the protective conductor PE, the RCD and/or the MCB are able to react—Fig. 5.3B. A fault current in the cases from Fig. 5.3A (parallel arc) and from Fig. 5.3B (earth/parallel arc) may reach values many times higher than the load current. Therefore parallel arcing gives a relatively favorable condition to be detected by the MCB (also by the RCD in the case of earth arc).

RCDs cannot be taken into account as a protection against fire in case of series arc faults. A series arc fault is formed only in one core (wire) of the conductor (Fig. 5.3C). The occurrence of such an arc is possible when a conductor is damaged due to its excessive bending, or during assembly and construction work, as well as due to rodents' activity. The value of current flowing through the faulty conductor (wire) is relatively low because the arc current is simultaneously the load current in this circuit.

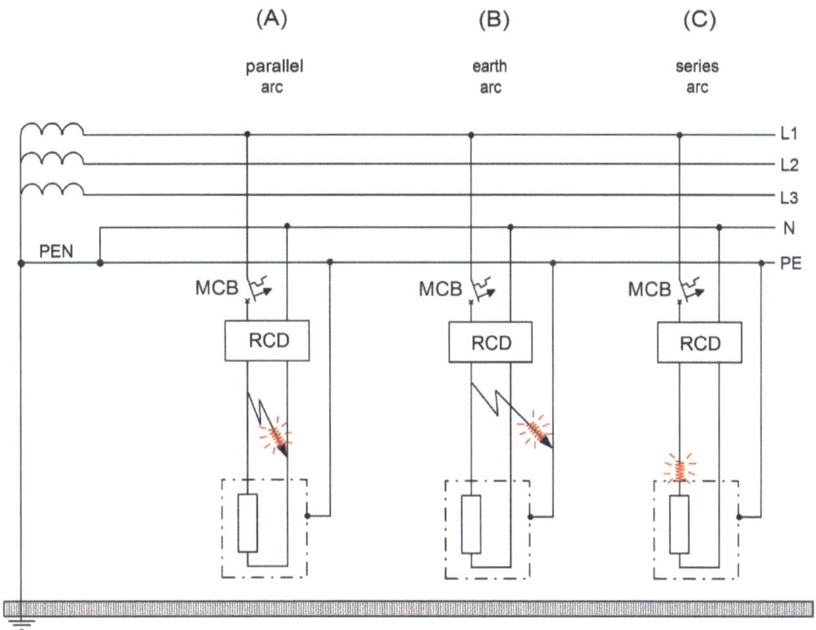

Fig. 5.3 Risk of fire due to the presence of: (A) parallel arc, (B) earth (parallel) arc, (C) series arc (arcing only in one core of the conductor); RCD—residual current device, MCB—miniature circuit-breaker.

Thus this fault is detected neither by residual current devices (there is no residual current) nor circuit-breakers (too low value of the flowing current). In order to detect insulation fault with arcing, especially when series arc fault occurs [4], arc fault detection devices (AFDD), constructed according to standard [3], are recommended to be used.

If mechanical stress is forced on the cable (e.g., by careless placement of the washing machine), the copper conductor of the cable can be narrowed (Fig. 5.4A) and it may be the starting point of the series arcing process. In the narrowed place, the density of the current is increased and in effect, the temperature of the conductor rises. The raised temperature makes oxidization of the conductor as well as carbonization of the insulation (Fig. 5.4B). These processes progress over time and the continuity of the conductor can be broken. In this step the series arcing starts along with the weak connection of both parts (sides) of the conductor (Fig. 5.4C). Series arcing inside the cable aggravates its degradation, so a fire may get outside the cable (Fig. 5.4D).

The development of the series arcing is clearly presented in the form of photos in the paper [5]. Fig. 5.5 presents photos from this paper.

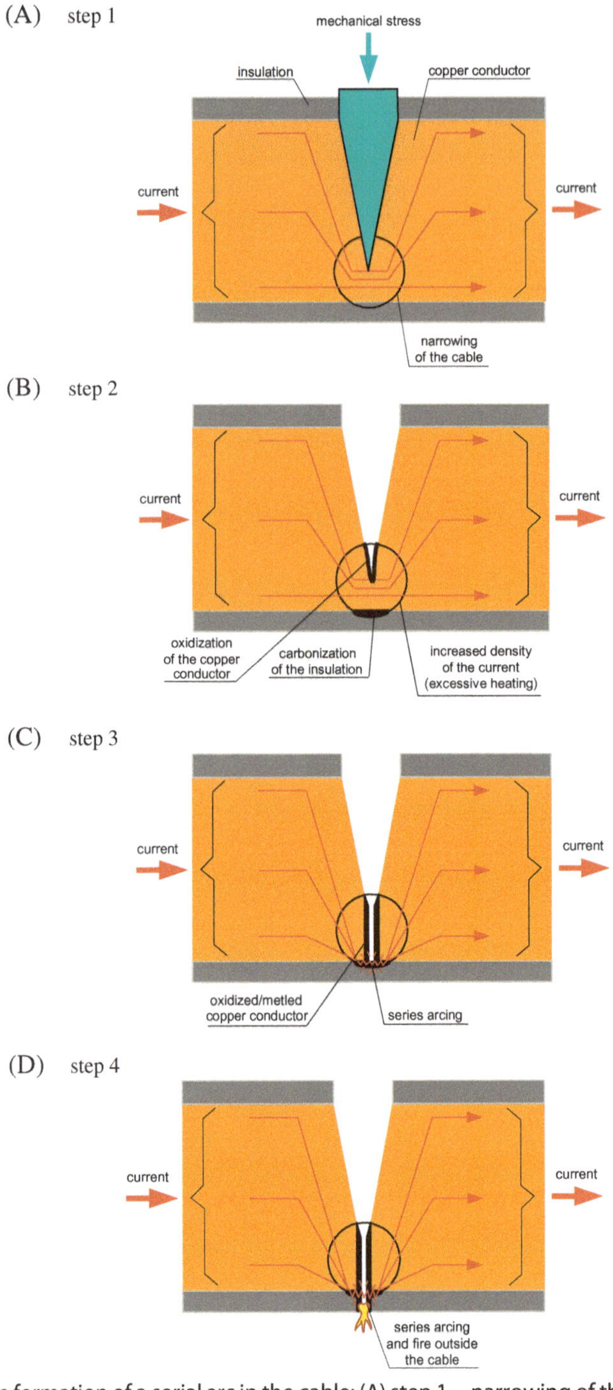

Fig. 5.4 The formation of a serial arc in the cable: (A) step 1—narrowing of the cable due to mechanical stress, (B) step 2—excessive heating of the narrowed conductor, (C) step 3—broken continuity of the conductor and starting the serial arcing, (D) step 4—serial arcing and fire outside the cable.

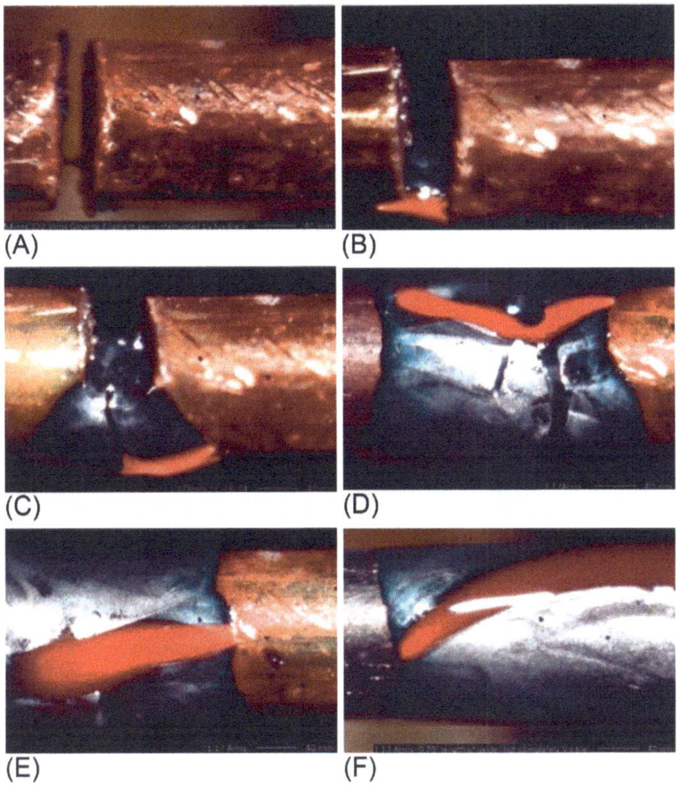

Fig. 5.5 Photographs showing the initial bridge formation, glowing filament (or "worm"), and growth progression of the copper oxide (dark area) between two 1 mm diameter, 37 × magnification, stationary copper wires at 1.17 A_{rms}. Exposure time was 125 ms. (A) Initial filament $t=0$ min, (B) oxide growth $t=2$ min, (C) $t=8$ min, (D) $t=28$ min, (E) $t=31$ min, (F) $t=64$ min [5].

To detect an arcing, the detection system of AFDDs cannot be based only on the peak or rms value of the current because, in the case of series arcing, the detected current can be lower than the maximal load current in the circuit. Hence, a system for detection and identification of arc fault included in AFDDs analyzes especially a *Received Signal Strength Indicator* (RSSI), which is utilized in telecommunications. It informs about the power present in a received radio signal. A simplified structure of the AFDD is presented in Fig. 5.6.

Comparison of this indicator: from arc fault vs. from load current of electric equipment (mainly high-frequency components of the analyzed currents) enables to distinguish a fault from a normal state [6]. Fig. 5.7

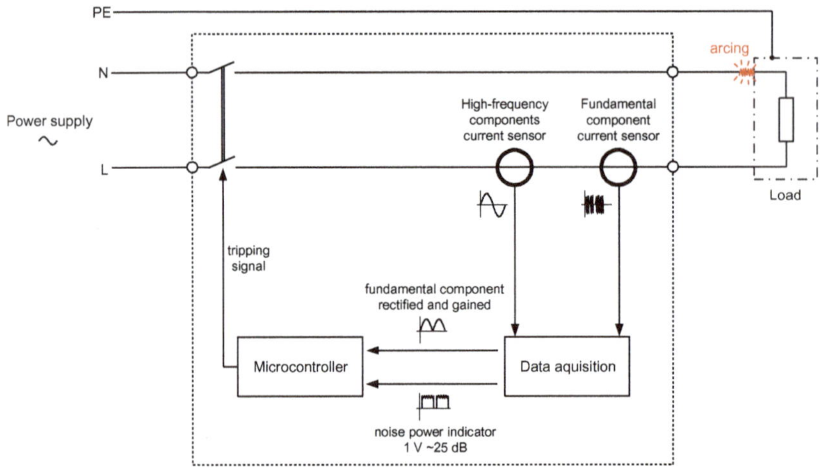

Fig. 5.6 Functional structure of the arc fault detection device (AFDD); according to [4].

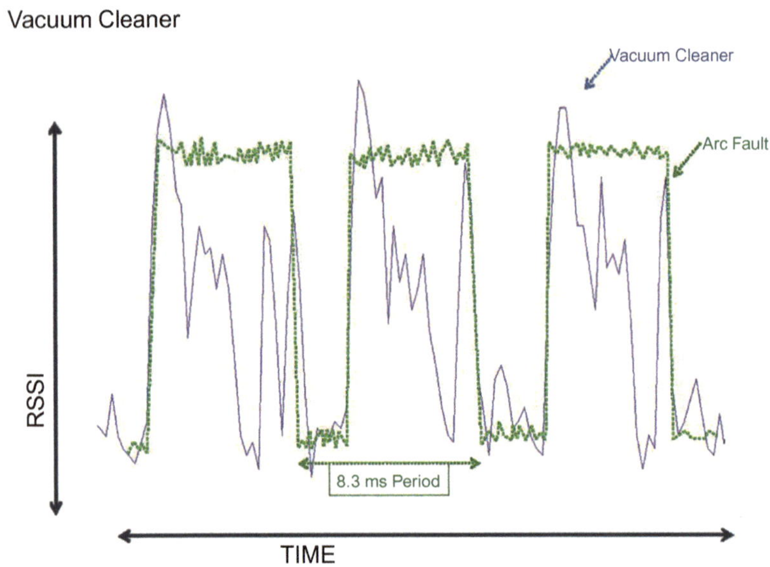

Fig. 5.7 Description of a vacuum cleaner load in comparison to an arc fault [7].

presents a comparison of the RSSI signal of an arc with the RSSI signal coming from a vacuum cleaner load current [7]. Similar comparisons for an electric drill and a home control signal are shown in Figs. 5.8 and 5.9, respectively [7].

Standard IEC 62606 [3] recognizes two types of tripping characteristics of AFDDs. When the test arc current is up to 63 A (low arc currents), the

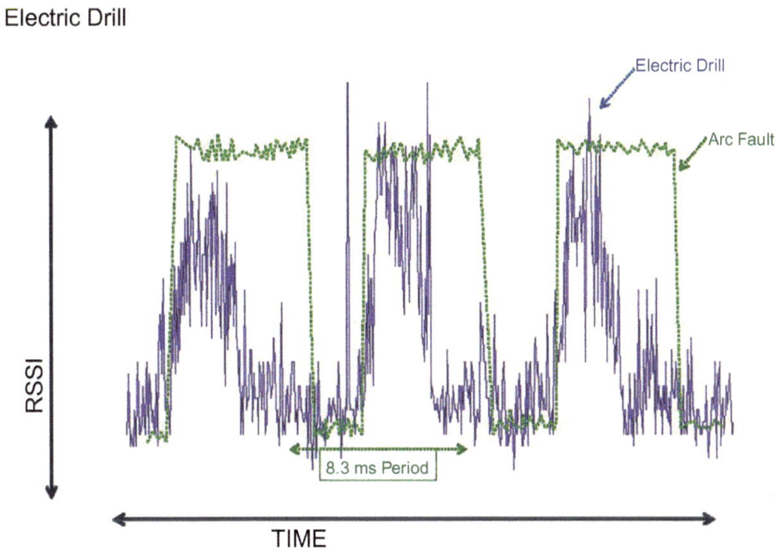

Fig. 5.8 Description of the broadband energy of an arc fault and of an electric drill [7].

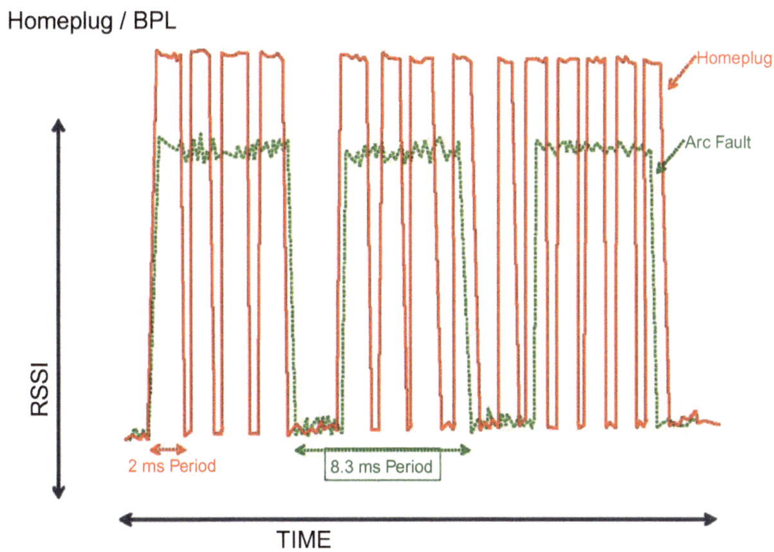

Fig. 5.9 Description of homeplug/broadband over the power line vs. that of a typical series arc fault [7].

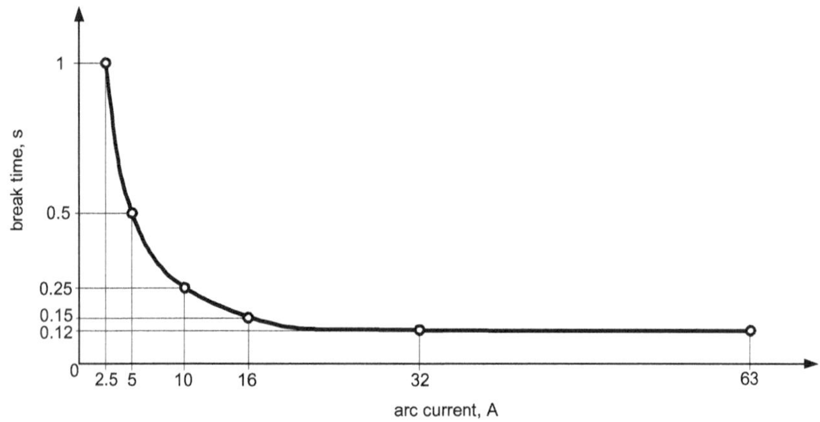

Fig. 5.10 Standard tripping characteristic of AFDDs for low arc current (rms; up to 63 A).

maximum break time as a function of the value of the arc current is specified (Fig. 5.10). For example, if the arc current is equal to 10 A, the time is 0.25 s. For the arc current within the range of 32–63 A, the maximum break time is 0.12 s. In the case of high arc current (more than 63 A), the standard [3] delivers another criterion—the maximum permissible number of half-cycles at the rated frequency as a function of prospective short-circuit current before arcing (Fig. 5.11). The maximum number of half-cycles is equal to 12 when the prospective short-circuit current before arcing is equal to 75. For the prospective short-circuit current within the range of 150–500 A, the maximum number of half-cycles is 8.

Fig. 5.12 presents a comparison of tripping characteristics of the RCD ($I_{\Delta n} = 300$ mA), AFDD, and the most popular MCBs used in final circuits of electrical installations (B16, C16, and D16). The highest/best sensitivity

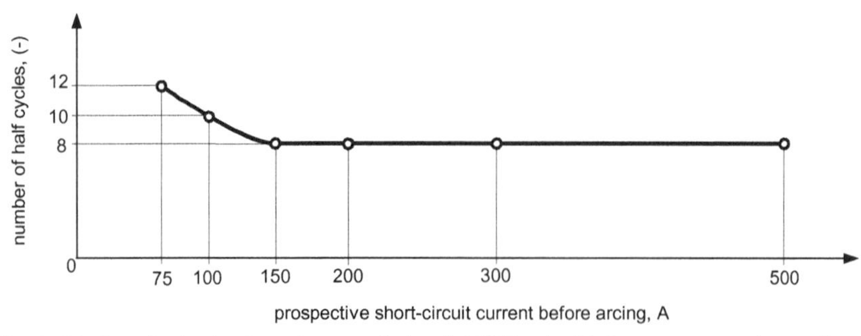

Fig. 5.11 Standard tripping characteristic of AFDDs for high arc current (rms; above 63 A).

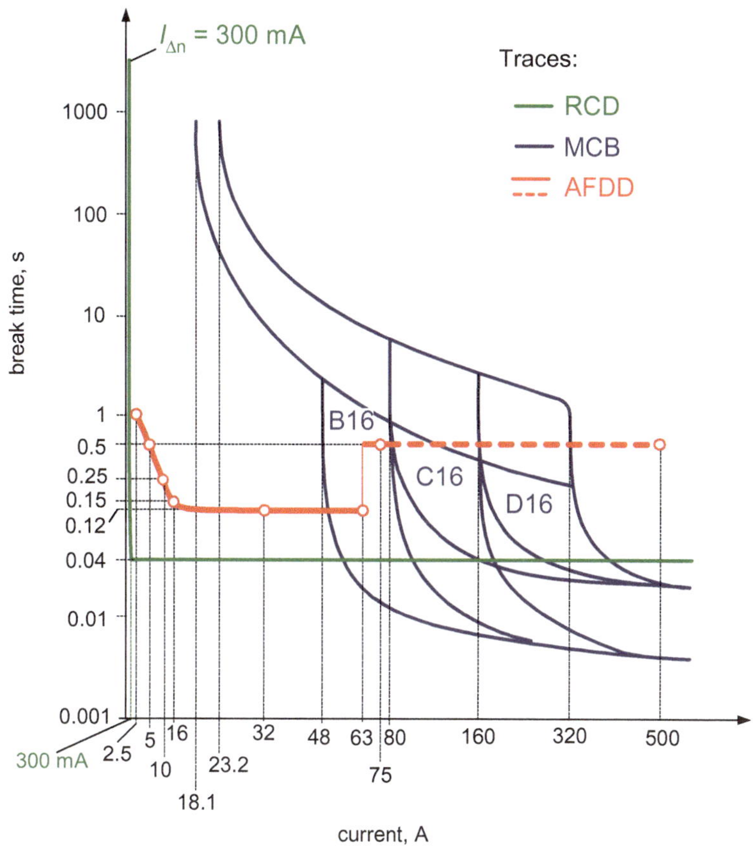

Fig. 5.12 Comparison of the tripping characteristics of the: RCD ($I_{\Delta n}$ = 300 mA), MCBs of I_n = 16 A (B16, C16, D16) and AFDD, as a function of the fault current in the electrical installation.

(300 mA) presents the RCD but one must be aware that RCDs can detect only residual (earth fault) currents, as in the case presented in Fig. 5.3B. Arcing currents less than 23.2 A (1.45 × 16—upper limit current of the MCB's thermal relay) are practically not detected by MCB, neither B16 nor C16 and D16. Fast reaction of the MCB may occur for currents at least: 80 A for B16, 160 A for C16, 320 A for D16. Therefore the aforementioned MCBs are not suitable for the detection of low arc currents (up to 63 A). The AFDD, performed according to standard IEC 62606 [3], enables for detection of arc currents within the range of 2.5–500 A, irrespective of what value of the rated current it has.

Arc fault detection devices should not cause false tripping—they should not react to the arc that occurs during the normal operation of electric equipment. This arc may appear during switching on lighting devices or electronic devices as well as appear in the case of the normal operation of motors included in handheld tools or domestic equipment (e.g., vacuum cleaner). The correct detection and identification of the arc generated by fault and the arc occurring during normal operation of the electric equipment is relatively difficult. It requires the application of an advanced system of arc parameters' identification. Therefore, as aforementioned, the decisive criterion for the tripping of the AFDD cannot only be the rms value or the amplitude of the load current because in the case of series arc fault these values are relatively low. They can be significantly lower compared to the load current in the circuit without fault.

The standard [3] provides the scope of the tests according to which AFDDs are obliged not to trip. These tests reflect the normal condition of the electric equipment operation in 230 V circuits and they are as follows:

- starting and operation of a vacuum cleaner of a rated current within the range 5–7 A,
- turning on an electronic switching mode power supply; a load current is at least 2.5 A; this power supply should have a minimum total harmonic distortion (THD) of 100% and individual minimum current harmonics of 75% at the 3rd, 50% at the 5th, and 25% at the 7th,
- starting and operation of a capacitor-start motor (air compressor type) of rated power of 2.2 kW,
- turning on a thyristor-type dimmer of the lamp (rated power 600 W) with filtering coil controlling a 600 W tungsten load; the dimmer should be turned on with the dimmer preset at full on, phase angles of 0°, 60°, 90°, 120°, and at the minimum setting that causes the lamps to ignite,
- turning on and operating two 40 W fluorescent lamps and an additional 5 A resistive load,
- turning on and operating of 12 V halogen lamps powered with an electronic transformer with a total power of at least 300 W and an additional 5 A resistive load,
- turning on and operation of an electric hand tool (e.g., a drill of at least 600 W power).

During these tests, five *Start-Stop* operations have to be performed and the loads have to be energized within the period of at least 5 s.

In the market offer, AFDDs are delivered as a separate protective device or can be integrated with an RCD and an MCB. The most important preferred and standard values, in terms of the selection of AFDDs, are presented as follows, according to the standard [3]:

- *Rated voltage* V_n:
  Preferred values: 230 V and 120 V (USA).
- *Rated current* I_n:
  Preferred values: 6–8–10–13–16–20–25–32–40–50–63 A.
- *Rated frequency f*:
  Preferred values: 50 and 60 Hz; if another value of the rated frequency is used, it shall be marked on the device. Relevant tests shall be carried out at this frequency.
- *The minimum value of the rated making and breaking capacity* I_m *and minimum value of the rated making and breaking capacity on one pole* I_{m1}:
  The minimum value of parameters I_m and I_{m1} is $10I_n$ or 500 A, whichever is greater. A similar requirement is referred to RCDs without integrated circuit-breakers. Thus AFDDs shall be backup protected.
- *Rated conditional short-circuit current* I_{nc} *and rated conditional short-circuit current for one pole* I_{nc1}:
  Standard values of these currents (values of AC component of a prospective current, which an AFDD, protected by a suitable short-circuit protective device in series, can withstand under the specified conditions) are 3, 4.5, 6, 10, 20, and 25 kA.

The application of AFDDs is recommended in some types of locations. The provisions referred to the AFDD application are included in standard HD 60364-4-42:2011/A1:2015 [8]. The protection with the use of these devices is preferred in final circuits supplying current-using equipment in the following places (Clause 421.7 of the standard HD 60364-4-42:2011/A1:2015):

- in premises dedicated for sleeping,
- in locations where increased risk of fire due to the nature of processed or stored materials exists (e.g., in barns, wood-working shops, or stores of combustible material),
- buildings which are made of combustible constructional materials, especially in wooden buildings,
- in building with increased risk of the fire propagating (e.g., very high buildings),
- in buildings/locations containing endangering or irreplaceable goods.

The aforementioned standard only recommends the application of the AFDD but in the USA they are required in certain types of low-voltage installations. In the USA these protective devices are called AFCIs (*arc fault circuit interrupter*), and according to National Electrical Code [9] they have to protect circuits (120 V, 15 A, and 20 A) supplying socket-outlets installed in dwelling unit family rooms, dining rooms, living rooms, bedrooms, sunrooms, recreation rooms, parlors, closets, hallways, libraries, or similar places.

In the USA, AFCIs have to be used in PV installations of a rated voltage of 80 V and above if the installations are integrated with a building [10]. They shall be suitable for the detection of arcing in DC circuits.

The requirement for the use of AFDDs was introduced in Germany in 2016 (but in 2019 the rules were liberalized). They were required in low-voltage circuits of a rated current up to 16 A, for supplying electric devices installed especially in bedrooms, daycare facilities for children, homes for older or disabled people, in locations with increased risks of fire (combustible building materials), and locations with irreplaceable goods—locations similar to those mentioned in standard HD 60364-4-42:2011/A1:2015 [8].

References

[1] HD 60364-4-42:2011, Low-voltage electrical installations—Part 4-42: Protection for safety—Protection against thermal effects, 2011.
[2] HD 60364-5-53:2015, Low-voltage electrical installations—Part 5-53: Selection and erection of electrical equipment—Switchgear and controlgear, 2015.
[3] IEC 62606:2013, General requirements for arc fault detection devices, 2013.
[4] OEZ Minia, Arc fault detection devices AFDD. Application manual, 2017.
[5] J.J. Shea, Conditions for series arcing phenomena in PVC wiring, IEEE Transactions on Components and Packaging Technologies 30 (3) (Sept. 2007).
[6] S. Czapp, Application of RCD and AFDD in low-voltage electrical installations for protection against fire, Przeglad Elektrotechniczny 95 (11) (2019) 12–16.
[7] C.E. Restrepo, Arc fault detection and discrimination methods, in: 53rd IEEE Holm Conference on Electrical Contacts, 2007, pp. 115–122.
[8] HD 60364-4-42:2011/A1:2015, Low voltage electrical installations – Part 4–42: Protection for safety – Protection against thermal effects, 2015.
[9] NEC, National Electrical Code (NEC), (2008).
[10] J. Johnson, C. Oberhauser, M. Montoya, A. Fresquez, S. Gonzalez, A. Patel, Crosstalk nuisance trip testing of photovoltaic DC arc-fault detectors, in: 38th IEEE Photovoltaic Specialists Conference, 2012, pp. 1383–1387.

CHAPTER 6

Requirements of standards with reference to application of residual current devices

Multipart standard IEC/HD 60364 dedicated to low-voltage electrical installations requires or recommends the application of RCDs in some types of installations or circuits. Provisions of particular parts of standard IEC/HD 60364 referring to the application of RCDs are compiled and presented as follows. For the more complicated or not quite clear cases, the author's comments related to the provisions of the standards have been added.

6.1 HD 60364-4-41:2017 Low-voltage electrical installations—Part 4-41: Protection for safety—Protection against electric shock [1]

Subclause 411.3.3 Further requirements for socket-outlets and for the supply of mobile equipment for use outdoors.

Additional protection with the use of RCDs with a rated residual operating current not exceeding 30 mA is required for:

- Socket-outlets (in AC circuits) with a rated current not exceeding 32 A, which are indented for general use, and can be used by ordinary persons, as well as
- AC mobile equipment with a rated current not exceeding 32 A, which can be used outdoors.

The aforementioned requirements do not apply for IT systems characterized by first-fault current not exceeding 15 mA.

Subclause 411.3.4 Additional requirements for circuits with luminaries in TN and TT systems.

Additional protection with the use of RCDs with a rated residual operating current not exceeding 30 mA is required in AC final circuits supplying luminaires in premises designed to accommodate a single household.

Residual Current Devices
https://doi.org/10.1016/B978-0-323-89783-9.00012-6

6.2 IEC 60364-5-53:2015 Low-voltage electrical installations—Part 5-53: Selection and erection of electrical equipment—Switchgear and controlgear [2]

Clause 532.2 Residual current devices for protection against the risk of fire.

In locations where a particular risk of fires exists, RCDs with a rated residual operating current not exceeding 300 mA are required to be used. They shall be installed at the origin of the protected circuit.

6.3 IEC 60364-5-55:2011 Electrical installations of buildings—Part 5-55: Selection and erection of electrical equipment—Other equipment [3]

Clause 559.8 Protection against electric shock for display stands for luminaires.

Circuits supplying such stands shall be protected alternatively by:

- SELV (Safety Extra-Low Voltage) /PELV (Protective Extra-Low Voltage), or,
- RCDs of $I_{\Delta n} \leq 30$ mA.

6.4 HD 60364-7-701:2007 Low-voltage electrical installations—Part 7-701: Requirements for special installations or locations—Locations containing a bath or shower [4]

Clause 701.415.1 Additional protection: residual current devices (RCD).

All circuits in locations equipped with a bath or shower shall be protected by one or more RCDs of $I_{\Delta n} \leq 30$ mA. This requirement does not apply to SELV and PELV circuits, and to circuits protected by the electrical separation of a single current-using equipment.

Comments:

Locations containing a bath or shower, due to their nature, are usually utilized in wet conditions. This is the cause of high risk of fatal electric shock. To limit the possibility of risky utilization of current-using equipment, such locations are divided into special zones. In an example bathroom (Fig. 6.1) the zones are as follows:

- zone 0: the interior of the bath,
- zone 1: the area outside the bath limited: horizontally—by the floor level and the height 225 cm as in Fig. 6.1; vertically—by the vertical plane surrounding the bath,
- zone 2: the parallel plane in the distance 60 cm (height 225 cm) from zone 1.

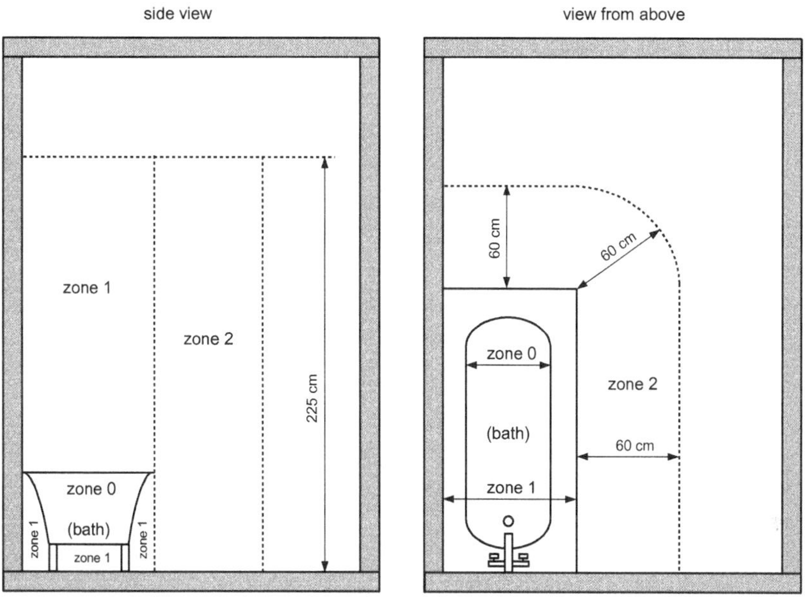

Fig. 6.1 Special zones in the example location with a bath.

In zone 0, only a power supply in the form of the SELV circuit, having nominal voltage not exceeding 12 V AC or 30 V DC, is permitted. For such a way of the circuit supply, RCDs are not necessary.

In zone 1, the following ways of the supply and protection of circuit are permitted:

- SELV circuit, having nominal voltage not exceeding 25 V AC or 60 V DC; if such a way of the circuit supply is applied, RCDs are not necessary,
- electrical separation for the supply of only single current-using equipment; if such a way of the circuit supply is applied, RCDs are not necessary,
- circuits other than SELV or other than electrical separation for the supply of only single current-using equipment; RCDs of $I_{\Delta n} \leq 30\,\text{mA}$ are necessary then.

6.5 IEC 60364-7-702:2010 Low-voltage electrical installations—Part 7-702: Requirements for special installations or locations—Swimming pools and fountains [5]

Provisions referring to the application of RCDs are included in *Clauses/Subclauses: 702.410.3.101.2, 702.410.3.101.3, 702.53, 702.55.101.3, 702.55.104.2, 702.55.105.*

After analysis of the aforementioned subclauses, it can be stated that in the case of protection by automatic disconnection of supply, RCDs of $I_{\Delta n} \leq 30\,\text{mA}$ shall be used.

Moreover, if a SELV power source is installed in zone 2 (for zones description, see *Comments*) to supply electric equipment in zone 0 or zone 1, this power source has to be protected by an RCD of $I_{\Delta n} \leq 30\,\text{mA}$. In some cases, the same requirement refers to the power source (transformer) of the electrical separation.

Comments:

Similarly to the location containing a bath or shower, in areas containing swimming pools or fountains special zones are also provided. For swimming pools, these zones include:

- zone 0: the interior of basins (also basins for foot cleaning), including any recessed in their walls or floors, the interior of waterjets or waterfalls and space below them,
- zone 1: the area outside the interior of basins limited: horizontally—the floor expected to be occupied by persons and the horizontal plane 2.5 m above this floor; vertically—by the plane 2 m measured from the rim of the basin; if the swimming pool contains starting blocks, springboards, chutes, or other components, which may be occupied by persons, the vertical plane 1.5 m and the horizontal plane 2.5 m, as in Fig. 6.2, should be taken into account.
- zone 2: the parallel plane in the distance of 1.5 m (height 2.5 m) from zone 1.

Fig. 6.2 presents an example swimming pool and special zones.

In some zones (e.g., zone 0 and zone 1 in swimming pools) current-using equipment is permitted to be supplied only from SELV circuits at a nominal voltage not exceeding 12 V AC or 30 V DC. For example, socket-outlets (only SELV) may be installed in zone 1 but the source of supply must be outside zone 0 and 1. If the source is located in zone 2, its supply shall be protected by an RCD of $I_{\Delta n} \leq 30\,\text{mA}$. As one can see, the level of safety is required to be very high.

6.6 HD 60364-7-703:2005 Electrical installations of buildings—Part 7-703: Requirements for special installations or locations—Rooms and cabins containing sauna heaters [6]

Clause 703.412.5.

In all circuits in rooms and cabins containing sauna heaters, except a circuit supplying a sauna heater, additional protection with the use of one or more RCDs of $I_{\Delta n} \leq 30\,\text{mA}$ shall be applied.

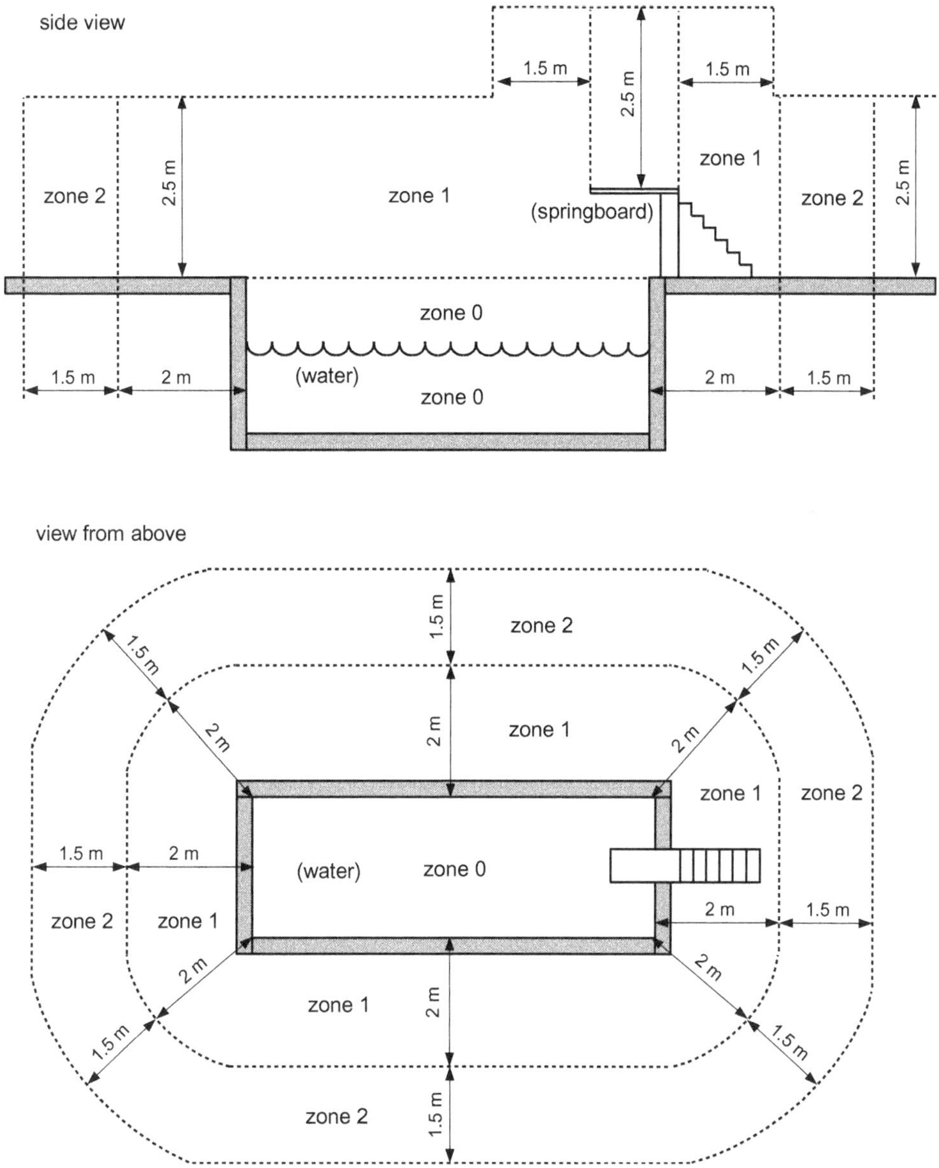

Fig. 6.2 Special zones in a swimming pool—an example.

6.7 HD 60364-7-704:2007 Low-voltage electrical installations—Part 7-704: Requirements for special installations or locations—Construction and demolition site installations [7]

Subclause 704.410.3.10.

In construction and demolition site installations, socket-outlet circuits with a rated current not exceeding 32 A and other circuits which supply hand-held equipment (also up to 32 A) shall be protected by RCDs of $I_{\Delta n} \leq 30\,\text{mA}$. This requirement does not apply to SELV and PELV as well as electrical separation for the supply of one item of current-using equipment.

Subclause 704.411.3.2.1.

Automatic disconnection of supply in socket-outlet circuits with a rated current exceeding 32 A shall be performed with the use of RCDs of $I_{\Delta n} \leq 500\,\text{mA}$.

6.8 HD 60364-7-705:2007 Low-voltage electrical installations—Part 7-705: Requirements for special installations or locations—Agricultural and horticultural premises [8]

Clause 705.411.1.

In agricultural and horticultural premises, for automatic disconnection of supply the following RCDs shall be used:

- $I_{\Delta n} \leq 30\,\text{mA}$—in socket-outlet final circuits with a rated current not exceeding 32 A,
- $I_{\Delta n} \leq 100\,\text{mA}$—in socket-outlet final circuits with a rated current exceeding 32 A,
- $I_{\Delta n} \leq 300\,\text{mA}$—in other circuits.

If high reliability of supply is very important, delay-type RCDs should be used.

6.9 HD 60364-7-706:2007 Low-voltage electrical installations—Part 7-706: Requirements for special installations or locations—Conducting locations with restricted movement [9]

Subclause 706.410.3.10.

In conducting locations with restricted movement, circuits supplying fixed equipment of class II shall have additional protection by the use of RCDs of $I_{\Delta n} \leq 30\,\text{mA}$.

Fig. 6.3 An example conducting location with restricted movement—deep trench with conductive walls and floor as well as restricted movement, in which persons are touching an earth potential and cannot free themselves from this potential.

Comments:

When a person uses electrical equipment in the conductive area and it is impossible to be free from the earth potential, there is an extremely high risk of electric shock. An example location with restricted movement is presented in Fig. 6.3. This is a deep trench having conductive walls and floor. For such a location special measures against electric shock have to be applied. In common electrical installations, the use of class II equipment is sufficient to fulfill the requirements of the standards in the aspect of protection in case of fault. However, in conducting locations with restricted movement it is not sufficient. Fixed equipment of class II shall have additional protection by the use of RCDs of $I_{\Delta n} \leq 30\,\text{mA}$.

6.10 IEC 60364-7-708:2017 Low-voltage electrical installations—Part 7-708: Requirements for special installations or locations—Caravan parks, camping parks and similar locations [10]

Clause 708.415.1.

In caravan parks, camping parks, and similar locations every socket-outlet shall be individually protected by an RCD of a rated residual operating current $I_{\Delta n} \leq 30\,\text{mA}$.

The same requirement (an RCD of $I_{\Delta n} \leq 30\,\text{mA}$) refers to a final circuit intended for the fixed connection for a supply to a mobile home or a residential park home.

6.11 HD 60364-7-709:2010/A12:2019 Low-voltage electrical installations—Part 7-709: Requirements for special installations or locations—Marinas and similar locations [11]

Clause 709.531.2.

In marinas and similar locations, every socket-outlet of a rated current not exceeding 63A shall be individually protected by an RCD of $I_{\Delta n} \leq 30\,\text{mA}$.

Every socket-outlet of a rated current exceeding 63A shall be individually protected by an RCD of $I_{\Delta n} \leq 300\,\text{mA}$.

When selectivity between RCDs in needed, S-type RCD is recommended to be used (upstream).

6.12 HD 60364-7-710:2012 Low-voltage electrical installations—Part 7-710: Requirements for special installations or locations—Medical locations [12]

Subclause 710.411.3.2.1.

In medical locations of group 1 and group 2, RCDs of type A or type B shall be used. It depends on the waveform shape of a possible residual current.

Clause 710.411.4, TN systems.

In medical locations of group 1, final circuits rated up to 32 A shall be protected by RCDs of $I_{\Delta n} \leq 30\,\text{mA}$.

In medical locations of group 2, automatic disconnection of supply with the use of RCDs with $I_{\Delta n} \leq 30\,\text{mA}$ shall only be applied to the following circuits:

- for the supply of operating tables,
- for X-ray units,
- for large equipment with a rated power greater than 5 kVA.

Clause 710.411.5, TT systems.

In medical locations of group 1 and group 2, the requirement of TN systems shall be fulfilled, and in all cases, RCDs are required to be used.

Subclause 710.62 g).

Tripping of RCDs under the current $I_{\Delta n}$ shall be verified not less than once per 12 months.

Comments:

Medical locations are divided into three groups (group 2, group 1, and group 0—according to HD 60364-7-710):

- Location of group 2—a medical location where applied parts are intended to be used in applications such as intracardiac procedures, operating theaters, and vital treatment where discontinuity (failure) of supply can cause danger to life; an intracardiac procedure is a procedure whereby an electrical conductor is placed within the heart of a patient or is likely to come into contact with the heart, such conductor being accessible outside the patient's body.
- Location of group 1—a medical location where applied parts are intended to be used externally invasively to any part of the body,
- Location of group 0—a medical location where no applied parts are intended to be used.

The applied part is part of the medical electrical equipment which in normal use comes into physical contact with the patient. For the equipment to perform its function, it must be in contact with the patient or needs to be touched by the patient.

In locations of groups 1 and 2, special medical equipment can be utilized. In circuits with such equipment, a nonsinusoidal earth fault current may flow. Therefore AC-type RCDs are not permitted there—the standard requires at least RCDs of type A.

The location of group 2 is the most demanding in terms of the reliability of supply. The high sensitivity RCDs ($I_{\Delta n} \leq 30\,\text{mA}$) may be the source of the nuisance tripping. Therefore, in TN systems their application is limited to the selected circuits of noncritical infrastructure (for the supply of operating tables, X-ray units, equipment with a rated power greater than 5 kVA).

6.13 IEC 60364-7-711:2018 Low-voltage electrical installations—Part 7-711: Requirements for special installations or locations—Exhibitions, shows and stands [13]

Subclause 711.410.3.101.

All socket-outlet final circuits of a rated current not exceeding 32 A, all final circuits supplying handheld electrical equipment of a rated current not exceeding 32 A, as well as final lighting circuits (except emergency lighting circuits), shall be protected by automatic disconnection of supply with the use of RCDs of a rated residual operating current $I_{\Delta n} \leq 30\,mA$. In such protected circuits, additional protection is also ensured.

Alternatively, SELV, PELV, or electrical separation of one item may be applied.

Subclause 711.411.3.2 Automatic disconnection in case of fault.

Subclause 711.411.3.2.101.

Automatic disconnection of supply with the use of RCDs shall be used in circuits supplying temporary structures.

Clause 711.535.3.

RCDs installed in series shall operate selectively.

6.14 HD 60364-7-712:2016 Low-voltage electrical installations—Part 7-712: Requirements for special installations or locations—Photovoltaic (PV) systems [14]

Subclause 712.530.3.101.

Where an RCD is used in a PV installation to protect the AC supply circuit, the B-type RCD has to be installed, unless:

- at least simple separation between the AC side and the DC side is ensured by the inverter, or
- at least simple separation between the AC side and the DC side is ensured by a transformer, or
- it is clearly indicated by the inverter's manufacturer that the inverter does not require a B-type RCD.

Comments:

Earth fault in a PV installation may have various waveform shapes. The waveform shape mainly depends on a point of earth fault (DC side or AC side), the structure of the PV inverter, as well as whether it contains a transformer. A general structure of the installation with PV energy sources is depicted in Fig. 6.4. In the installation without a simple separation between the AC side and the DC side, in the case of the earth fault at the DC side, the earth current is composed of AC waveform and DC component. The composite waveform can be various [15]. Fig. 6.5 presents example waveforms of the earth current (high-frequency components generated by the inverter switching frequency are neglected).

Since the earth current can be unidirectional of the low pulsation, only RCD of B-type is suitable. At this point it has to be mentioned that, generally, the RCD is not required in PV installations. However, if the installation designer has decided to apply an RCD, the aforementioned requirements (referred to the RCD type) are required to be fulfilled.

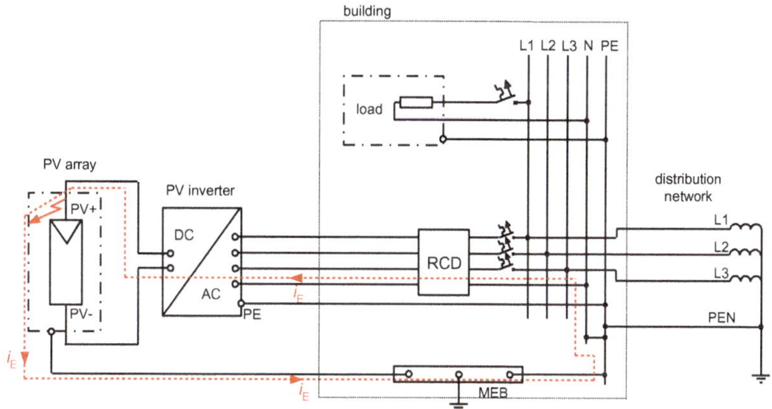

Fig. 6.4 A simplified structure of the installation with PV sources of energy; RCD—residual current device, MEB—main equipotential bonding busbar.

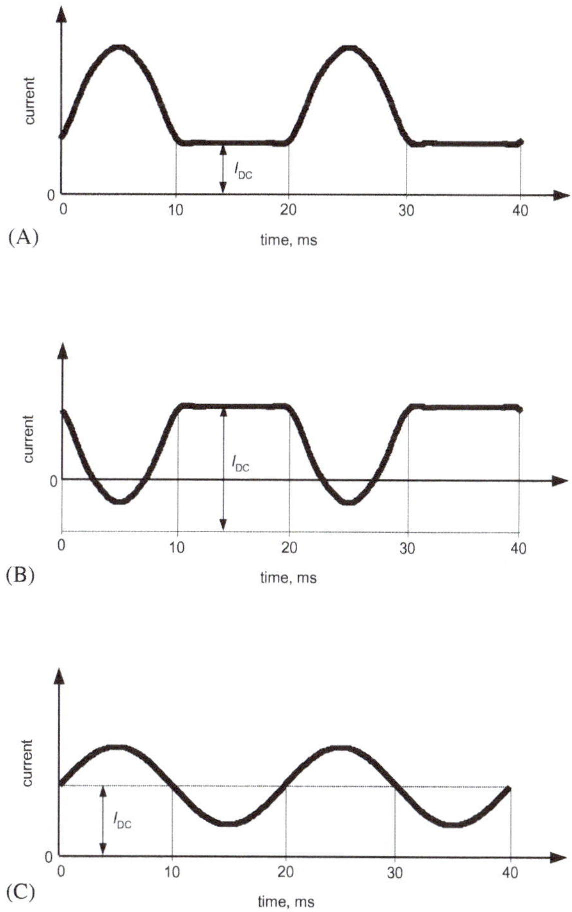

Fig. 6.5 Approximate example earth fault current waveforms in the case of the earth fault at DC side of the PV installation: (A) pulsating half-wave (positive) superimposed by smooth DC component, (B) pulsating half-wave (negative) superimposed by smooth DC component, (C) AC waveform superimposed by smooth DC component; I_{DC}—smooth DC component.

6.15 IEC 60364-7-713:2013 Electrical installations of buildings—Part 7-713: Requirements for special installations or locations—Furniture [16]

Subclause 713.415.1.1.

The electrical installation of the furniture shall have additional protection by an RCD with a rated residual operating current $I_{\Delta n} \leq 30\,\text{mA}$.

6.16 HD 60364-7-714:2012 Low-voltage electrical installations—Part 7-714: Requirements for special installations or locations—External lighting installations [17]

Subclause 714.411.3.3.

Lighting equipment installed in telephone kiosks, bus shelters, advertising panels, town plans, and similar installations shall be provided with additional protection by an RCD with a rated residual operating current $I_{\Delta n} \leq 30\,\text{mA}$.

6.17 HD 60364-7-721:2019 Low-voltage electrical installations—Part 7-721: Requirements for special installations or locations—Electrical installations in caravans and motor caravans [18]

Clause 721.415.101 Additional protection: residual current protective devices (RCDs).

In electrical installations in caravans and motor caravans, for protection by automatic disconnection of supply an RCD with a rated residual operating current $I_{\Delta n} \leq 30\,\text{mA}$ shall be used.

Every supply inlet shall be directly connected to its own RCD.

During the selection of RCDs, special attention should be given to the characteristics of the load.

6.18 HD 60364-7-722:2018 Low-voltage electrical installations—Part 7-722: Requirements for special installations or locations—Supplies for electric vehicles [19]

Subclause 722.411.3.3, Subclause 722.531.2.101.

Every connection point for electric vehicles' supply shall be protected by its own RCD (at least type A) having a rated residual operating current $I_{\Delta n} \leq 30\,\text{mA}$. This requirement does not apply to circuits using the protective measure of electrical separation.

In a supply circuit equipped with a socket-outlet or vehicle connector complying with the EN 62196 series, protection against DC earth fault current shall be provided (except where provided by the EV charging station), as follows:

- RCD type B or,
- RCD type A and residual direct current detection device (RDC-DD),
- RCD type F and residual direct current detection device (RDC-DD).

Comments:

There are four modes of electric vehicles charging [19,20]:

- mode 1—the electric vehicle is connected to the AC supply network; the connection utilizes standard (as in residential buildings) socket-outlets not exceeding 16 A (250 V single-phase or 480 V three-phase), live conductors, and the protective earth conductor (Fig. 6.6A);
- mode 2—the electric vehicle is connected to the AC supply network; the connection utilizes standard (as in residential buildings) socket-outlets not exceeding 32 A (250 V single-phase or 480 V three-phase), live conductors, and the protective earth conductor together with a control pilot function (communications) and system of protection against electric shock based on an RCD between the electric vehicle and the plug or as part of the in-cable control protective device—IC-CPD (Fig. 6.6B);
- mode 3—the electric vehicle is connected to the AC supply network with the use of dedicated electric vehicle supply equipment (EVSE); control and protection functions (communication) are installed permanently in the installation (Fig. 6.6C);
- mode 4—the electric vehicle is connected to the AC supply network with the use of an off-board charger; control and protection functions (communication) as well as the vehicle charging cable are installed permanently in the installation (Fig. 6.6D).

Every electric vehicle charging circuit should be equipped with a standard RCD or other protective devices which ensure protection against residual currents. One of the most important problems in EV charging circuits is the occurrence of the DC component in the earth current. The value of this component may be too high for proper operation of even A-type and F-type RCDs. For such a case, as a supplementary device to the RCD, a residual direct current detection device (RDC-DD) should be used to detect a DC component higher than 6 mA (10 mA in the case of F-type RCDs). The RDC-DD cooperates with an external switch to open the circuit. Alternatively, an RCD of B-type can be applied.

Details for RDC-DD are included in standard IEC 62955:2018 "Residual direct current detecting device (RDC-DD) to be used for mode 3 charging of electric vehicles."

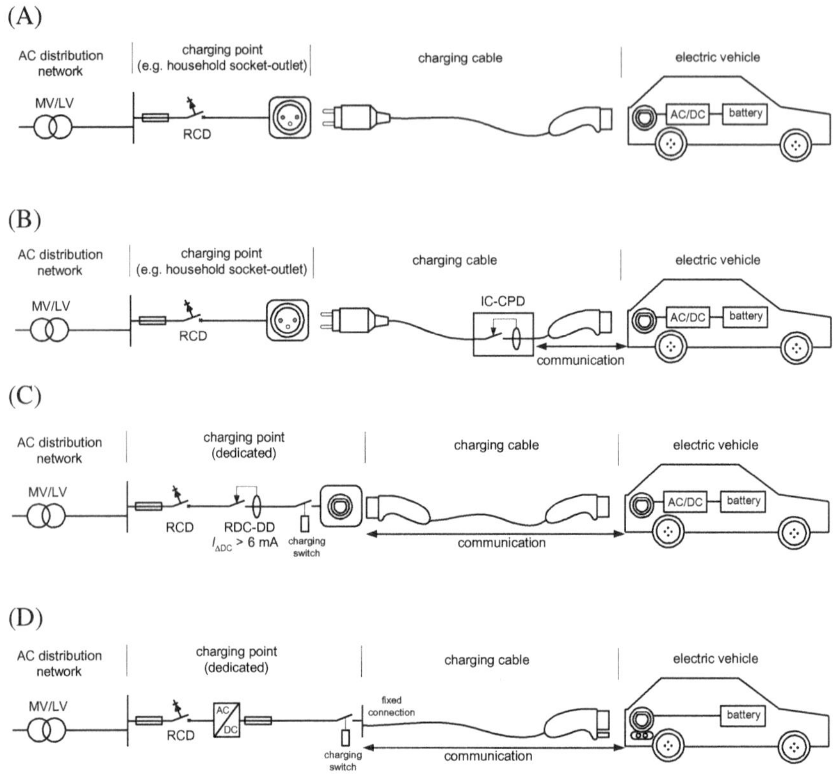

Fig. 6.6 Electric vehicle charging modes: (A) mode 1, (B) mode 2, (C) mode 3, (D) mode 4.

6.19 HD 60364-7-730:2015 Low-voltage electrical installations—Part 7-730: Requirements for special installations or locations—Onshore units of electrical shore connections for inland navigation vessels [21]

Clause 730.531.2.

In installations of onshore units of electrical shore connections for inland navigation vessels, when protection against indirect contact by automatic disconnection of supply is used, socket-outlets with a rated current up to 63 A shall be individually protected by an RCD of $I_{\Delta n} \leq 30\,\text{mA}$. For socket-outlets circuits with a rated current above 63 A, an RCD of $I_{\Delta n} \leq 300\,\text{mA}$ shall be used.

6.20 HD 60364-7-740:2006 Electrical installations of buildings—Part 7-740: Requirements for special installations or locations—Temporary electrical installations for structures, amusement devices and booths at fairgrounds, amusement parks and circuses [22]

Clause 740.412.5.

In temporary electrical installations for structures, amusement devices and booths at fairgrounds, amusement parks, and circuses, all lighting circuits (except emergency lighting circuits), socket-outlet final circuits with a rated current not exceeding 32 A, and circuits intended for supply of portable equipment connected by flexible cable or cord with a rated current not exceeding 32A shall be protected by RCDs with a rated residual operating current $I_{\Delta n} \leq 30\,\text{mA}$. This requirement does not apply to socket-outlet circuits using the protective measure of electrical separation or SELV/PELV as well as lighting circuits placed out of arm's reach not supplied by socket-outlets for household and similar purposes or socket-outlets in accordance with EN 60309-1.

Subclause 740.481.3.1.3.

Protection by automatic disconnection of supply of temporary installations shall occur in the origin of the installation by RCDs of $I_{\Delta n} \leq 300\,\text{mA}$. For this purpose, RCDs with the delay shall be applied.

6.21 IEC 60364-7-753:2014 Low-voltage electrical installations—Part 7-753: Requirements for special installations or locations—Heating cables and embedded heating systems [23]

Subclause 753.415.1.1.

In installations with heating cables and embedded heating systems, circuits supplying heating units shall be protected by RCDs with a rated residual operating current $I_{\Delta n} \leq 30\,\text{mA}$. The use of RCDs with the delay is not permitted.

Important notice:

The above-mentioned provisions of standard IEC/HD 60364 clearly show that RCDs are absolutely required in many types of circuits. In particular, it is required to install high sensitivity RCDs ($I_{\Delta n} \leq 30\,\text{mA}$). However, it should be noted that such RCDs have "high sensitivity" for natural leakage currents, even in situations when no risk of electric shock exists. This high sensitivity may result in the unwanted tripping of RCDs. There are critical electrical installations in which continuity of supply is of paramount importance, where there is no permission to the unwanted tripping. It refers, for example, to safety services electrical installations covered by standard IEC 60364-5-56. Following, the example provisions referring to nonrecommended or excluded RCDs use are presented.

6.22 IEC 60364-5-56:2018 Low-voltage electrical installations—Part 5-56: Selection and erection of electrical equipment—Safety services [24]

Subclause 560.7.13.

Safety services circuits shall not be protected by RCDs or AFDDs.

Comments:

RCDs are not permitted in safety services circuits, e.g., in emergency lighting circuits.

6.23 HD 60364-7-710:2012 Low-voltage electrical installations—Part 7-710: Requirements for special installations or locations—Medical locations [12]

Subclause 710.411.3.2.1.

The simultaneous use of many items of current-using equipment connected to the circuit cannot cause the unwanted tripping of the RCD.

Comments:

This provision signalizes that leakage currents (in a transient or in a steady state) may cause the unwanted tripping of an RCD protecting many items of current-using equipment. It is very undesirable in medical locations, in which specialized receivers may be installed/supplied. Therefore, in the installation with such equipment, a sufficient number of circuits, to avoid supplying too many items of current-using equipment from one circuit protected by an RCD, should be provided.

6.24 HD 60364-7-718:2013 Low-voltage electrical installations—Part 7-718: Requirements for special installations or locations—Communal facilities and workplaces [25]

Subclause 718.559.101.1.

If RCDs are used, the particular RCD can protect only one lighting circuit.

Comments:

Firstly, it should be noted that RCDs are not required in the lighting circuit of communal facilities and workplaces. However, if it is decided to use them, then the following remarks should be taken into account.

Communal facilities and workplaces usually gather a significant number of people. In the case of the sudden turn off of the light in a large hall (e.g., in an auditorium hall without a natural daylight) it may lead to panic. Reckless application of the RCD—one RCD protecting a few circuits, as in Fig. 6.7A—may result in nuisance tripping of this RCD under a transient

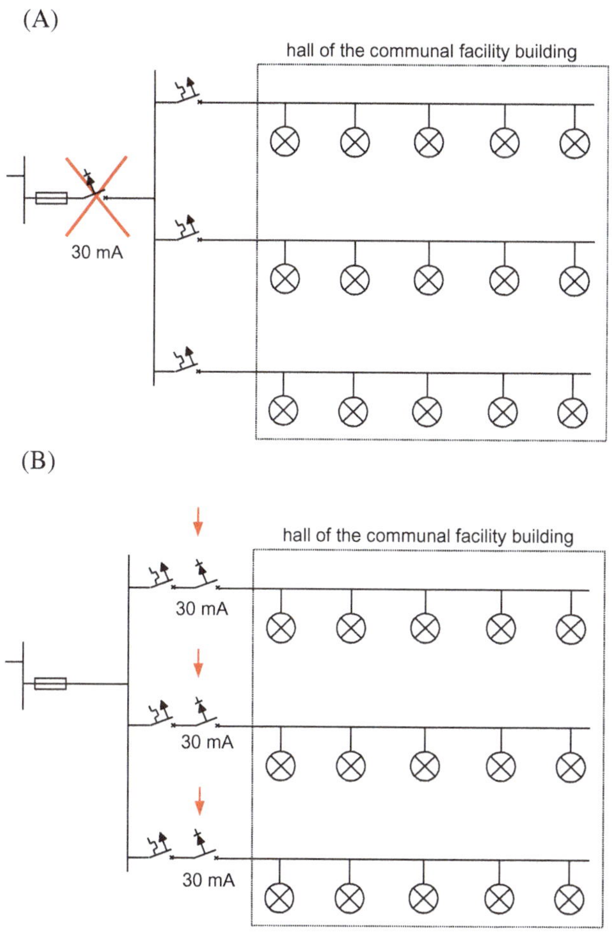

Fig. 6.7 Application of RCDs in lighting circuits of a communal facility: (A) improper—one RCD protects three lighting circuits, (B) proper—one RCD protects only one lighting circuit.

leakage current. In consequence, all lighting circuits are without a power supply. To avoid this uncomfortable and even risky incident, RCDs should be installed in final circuits individually, as in Fig. 6.7B.

References

[1] HD 60364-4-41:2017, Low-voltage electrical installations—Part 4-41: Protection for safety—Protection against electric shock, 2017.
[2] IEC 60364-5-53:2015, Low-voltage electrical installations—Part 5-53: Selection and erection of electrical equipment—Switchgear and controlgear, 2015.
[3] IEC 60364-5-55:2011, Electrical installations of buildings—Part 5-55: Selection and erection of electrical equipment—Other equipment, 2011.
[4] HD 60364-7-701:2007, Low-voltage electrical installations—Part 7-701: Requirements for special installations or locations—Locations containing a bath or shower, 2007.
[5] IEC 60364-7-702:2010, Low-voltage electrical installations—Part 7-702: Requirements for special installations or locations—Swimming pools and fountains, 2010.
[6] HD 60364-7-703:2005, Electrical installations of buildings—Part 7-703: Requirements for special installations or locations—Rooms and cabins containing sauna heaters, 2005.
[7] HD 60364-7-704:2007, Low-voltage electrical installations—Part 7-704: Requirements for special installations or locations—Construction and demolition site installations, 2007.
[8] HD 60364-7-705:2007, Low-voltage electrical installations—Part 7-705: Requirements for special installations or locations—Agricultural and horticultural premises, 2007.
[9] HD 60364-7-706:2007, Low-voltage electrical installations—Part 7-706: Requirements for special installations or locations—Conducting locations with restricted movement, 2007.
[10] IEC 60364-7-708:2017, Low-voltage electrical installations—Part 7-708: Requirements for special installations or locations—Caravan parks, camping parks and similar locations, 2017.
[11] HD 60364-7-709:2010/A12:2019, Low-voltage electrical installations—Part 7-709: Requirements for special installations or locations—Marinas and similar locations, 2019.
[12] HD 60364-7-710:2012, Low-voltage electrical installations—Part 7-710: Requirements for special installations or locations—Medical locations, 2012.
[13] IEC 60364-7-711:2018, Low-voltage electrical installations—Part 7-711: Requirements for special installations or locations—Exhibitions, shows and stands, 2018.
[14] HD 60364-7-712:2016, Low-voltage electrical installations—Part 7-712: Requirements for special installations or locations—Photovoltaic (PV) systems, 2016.
[15] Davids S., Grünebast G.: Residual Currents in Photovoltaic Installations, Version 1.1, 2011, Doepke Schaltgeräte GmbH.
[16] IEC 60364-7-713:2013, Electrical installations of buildings—Part 7-713: Requirements for special installations or locations—Furniture, 2013.
[17] HD 60364-7-714:2012, Low-voltage electrical installations—Part 7-714: Requirements for special installations or locations—External lighting installations, 2012.
[18] HD 60364-7-721:2019, Low-voltage electrical installations—Part 7-721: Requirements for special installations or locations—Electrical installations in caravans and motor caravans, 2019.
[19] HD 60364-7-722:2018, Low-voltage electrical installations—Part 7-722: Requirements for special installations or locations—Supplies for electric vehicles, 2018.

[20] IEC 61851-1:2017 Electric vehicle conductive charging system—Part 1: General requirements.
[21] HD 60364-7-730:2015, Low-voltage electrical installations—Part 7-730: Requirements for special installations or locations—Onshore units of electrical shore connections for inland navigation vessels, 2015.
[22] HD 60364-7-740:2006, Electrical installations of buildings—Part 7-740: Requirements for special installations or locations—Temporary electrical installations for structures, amusement devices and booths at fairgrounds, amusement parks and circuses, 2006.
[23] IEC 60364-7-753:2014, Low-voltage electrical installations—Part 7-753: Requirements for special installations or locations—Heating cables and embedded heating systems, 2014.
[24] IEC 60364-5-56:2018, Low-voltage electrical installations—Part 5-56: Selection and erection of electrical equipment—Safety services, 2018.
[25] HD 60364-7-718:2013, Low-voltage electrical installations—Part 7-718: Requirements for special installations or locations—Communal facilities and workplaces, 2013.

CHAPTER 7

Backup protection of residual current devices

7.1 General information

In case of a short-circuit in an electrical installation, between line conductors (L–L) or line conductor and the neutral conductor (L–N), high value of current may flow through an RCD. Such type of short-circuit does not trip the RCD but may generate high electrodynamic and thermal stress to it. RCDs have to withstand these stresses in the closed position.

International standard IEC 61008-1 [1] requires that residual current devices of RCCB type (without integral overcurrent protection) withstand values of peak current $I_{\text{p-k}}$ and Joule integral I^2t included in Table 7.1.

Peak current is responsible for the electrodynamic stress, and Joule integral for the thermal stress. If the short-circuit current gives higher values of peak current $I_{\text{p-k}}$ and Joule integral I^2t than presented in Table 7.1, an RCCB shall be backup protected. As a backup protection device, a fuse is recommended to be used, according to Table 3.3. In most cases, RCCBs should be backup protected by a gG-type fuse of a rated current not exceeding 63 A (graphical symbol 6000). However, in practice, RCCBs are very often in series with miniature circuit-breakers (MCBs of B-type, C-type, or D-type) instead of the above-mentioned fuse. In such a case, coordination between an RCCB and an MCB has to be verified.

7.2 Backup protection with the use of fuses

If an RCCB is marked 6000, it means that the highest permissible value of the rated current of a fuse installed in series with the RCCB is gG63 A. Such a composition of the protective devices (gG fuse and RCCB) ensures the safe breaking of short-circuit currents up to 6 kA. It also means that the RCCB withstands in closed position prospective currents up to 6 kA.

From the above-mentioned marking, one can find the maximal peak current and Joule integral which may appear in the analyzed circuit. According to

https://doi.org/10.1016/B978-0-323-89783-9.00014-X

Table 7.1 Values of peak current I_{p-k} and Joule integral I^2t which RCCBs have to withstand; according to [1].

I_{nc}, $I_{\Delta c}$ A		I_n, A								
		≤ 16	≤ 20	≤ 25	≤ 32	≤ 40	≤ 63	≤ 80	≤ 100	≤ 125
500	I_{p-k}, kA	0.45	0.47	0.5	0.57					
	I^2t, kA^{2}s	0.4	0.45	0.53	0.68					
1000	I_{p-k}, kA	0.65	0.75	0.9	1.18					
	I^2t, kA^{2}s	0.50	0.9	1.5	2.7					
1500	I_{p-k}, kA	1.02	1.1	1.25	1.5	1.9	2.1			
	I^2t, kA^{2}s	1	1.5	2.4	4.1	9.75	22			
3000	I_{p-k}, kA	1.1	1.2	1.4	1.85	2.35	3.3	3.5	3.8	3.95
	I^2t, kA^{2}s	1.2	1.8	2.7	4.5	8.7	22.5	26	42	72.5
4500	I_{p-k}, kA	1.15	1.3	1.5	2.05	2.7	3.9	4.3	4.8	5.6
	I^2t, kA^{2}s	1.45	2.1	3.1	5.0	9.7	28	31	45	82.0
6000	I_{p-k}, kA	1.3	1.4	1.7	2.3	3	4.05	4.7	5.3	5.8
	I^2t, kA^{2}s	1.6	2.4	3.7	6.0	11.5	25	31	48	65.0
10,000	I_{p-k}, kA	1.45	1.8	2.2	2.6	3.4	4.3	5.1	6	6.4
	I^2t, kA^{2}s	1.9	2.7	4	6.5	12	24	31	48	60.0

I_n, rated current; I_{nc}, rated conditional short-circuit current; $I_{\Delta c}$, rated conditional residual short-circuit current; I_{p-k}, peak current; I^2t, Joule integral.

standard IEC 60269-2 [2], in case of a short-circuit, the maximal let-through energy (Joule integral) of gG63 is $I^2t_{gG63} = 21{,}200\,A^2s$. For the example fuse presented in Fig. 7.1A [3], value of I^2t_{gG63} is lower than the permissible $21{,}200\,A^2s$, and approx. constant for short-circuit currents $I_k \geq 2.3$ kA, due to the current waveform cutoff phenomena. This is the first important factor (let-through energy). The second important factor is the maximal value of the peak short-circuit current in the circuit with gG63. Fig. 7.1B presents cutoff characteristic of gG fuses with a marked level of the cutoff current value i_o for the gG63 fuse ($i_o = 4.2$ kA).

Taking the earlier mentioned into account, RCCBs marked  6000 are able to withstand the following quantities:

- Joule integral $I^2t = 21{,}200\,A^2s$ (value from IEC 60269-2 [2]),
- peak current $I_{p\text{-}k} = i_o = 4.2$ kA (cutoff current for value of a prospective short-circuit current $I_k = 6$ kA; derived from Fig. 7.1B).

These values should be adopted as a reference. This assumption is very important in case of consideration of the use of other types of the fuse or the same type of the fuse but of other rated currents. Table 7.2 presents values of let-through energy and cutoff current for selected fuses in case of prospective short-circuit current $I_k = 6$ kA. Bold font presents maximal permissible value, but italic font informs that the permissible value is exceeded.

One can see that the maximal permissible rated current of a fuse is $I_n = 63$ A—both gG and gF types (permissible peak current $I_{p\text{-}k}$ and Joule integral I^2t are not exceeded). However, when short-circuit current value is lower than 6 kA, cutoff current of fuses decreases (for gG fuses, see Fig. 7.1B), and it gives the possibility to use a fuse of higher rated current. Comparison of characteristic parameters of fuses for prospective short-circuit current $I_k = 4$ kA is presented in Table 7.3. In this case, the highest permissible value of the rated current of a fuse is $I_n = 80$ A (instead of 63 A)—both gG and gF types. Note, that for prospective short-circuit current $I_k = 4$ kA the let-though energy remains the same as for $I_k = 6$ kA, only cutoff current values have been decreased.

7.3 Backup protection with the use of miniature circuit-breakers

RCCBs are very often installed in series with miniature circuit-breakers (MCBs). Thermal and electrodynamic stresses are usually higher than in

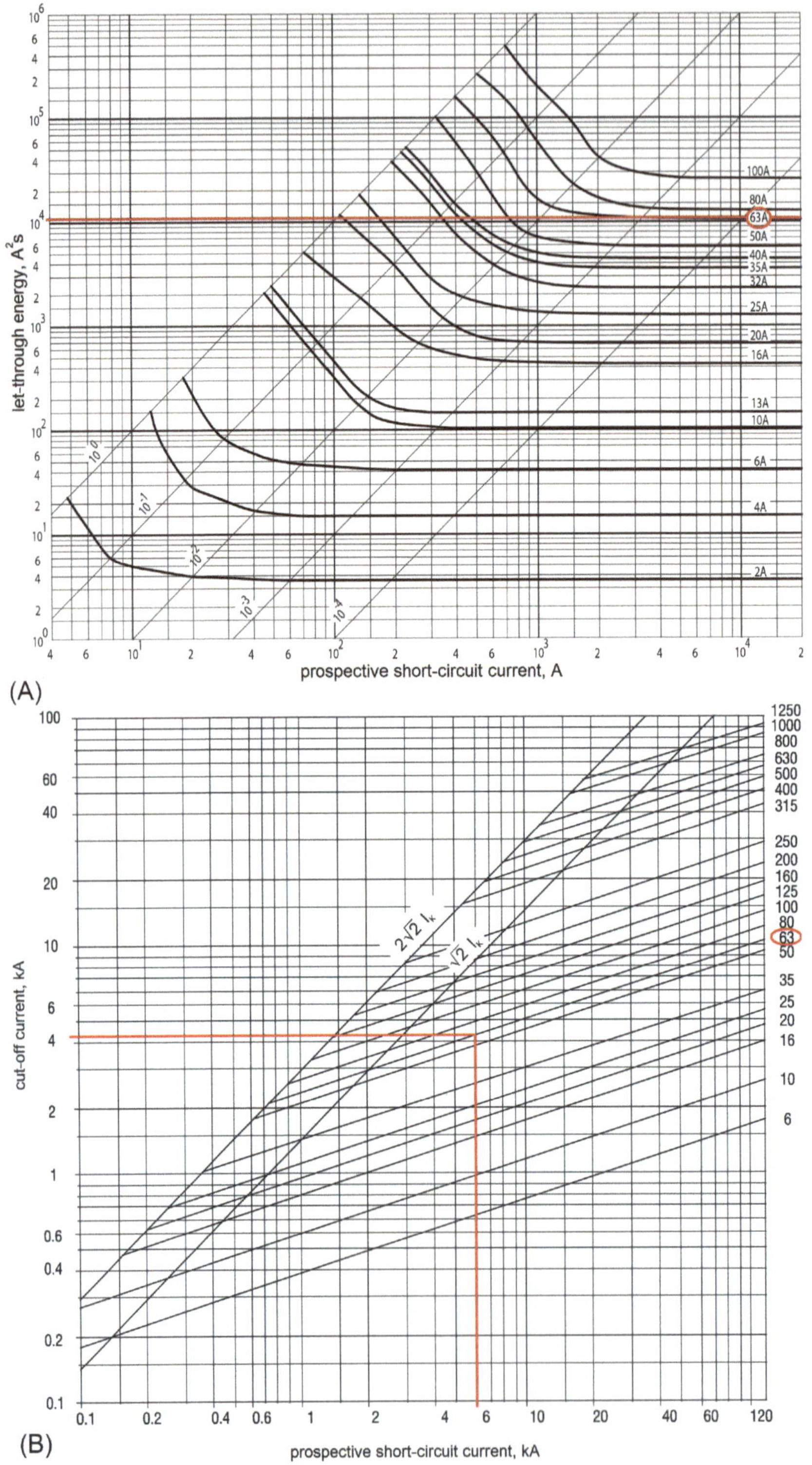

Fig. 7.1 Let-through energy (A) and cutoff current (B) of gG fuses [3], with marked example parameters for gG63.

Table 7.2 Characteristic parameters of fuses; prospective short-circuit current $I_k = 6\,kA$; data according to [2,3].

Rated current of a fuse A	Let-through energy A^2s gG (General type)	Let-through energy A^2s gF[a] (Fast type)	Cutoff current kA gG (General type)	Cutoff current kA gF[a] (Fast type)	Backup protection
25	4000	1210	2.1	1.7	Acceptable
50	13,700	5750	3.8	2.9	
63	**21,200**	9000	**4.2**	3.5	
80	*36,000*	13,700	*5.0*	*4.3*	Not acceptable
100	*64,000*	**21,200**	*5.9*	*5.5*	

Bold font—max permissible values (reference parameters for gG63).
[a]Not standardized by the IEC, standardized only by Polish local regulation.

comparison to the case with backup protection by fuses. For the first, when short-circuit current increases, let-through energy (Joule integral) increases as it is presented in Fig. 7.2A. In case of fuses it is constant for high-value short-circuit currents.

For the second, also cutoff current rises with short-circuit current rising (Fig. 7.2B). Moreover, many manufacturers do not deliver data concerning this cutoff current for miniature circuit-breakers, which makes analysis difficult.

Table 7.3 Characteristic parameters of fuses; prospective short-circuit current $I_k = 4\,kA$; data according to [2,3].

Rated current of a fuse A	Let-through energy A^2s gG (General type)	Let-through energy A^2s gF[a] (Fast type)	Cutoff current kA gG (General type)	Cutoff current kA gF[a] (Fast type)	Backup protection
25	4000	1210	1.8	1.4	Acceptable
50	13,700	5750	3.2	2.5	
63	**21,200**	9000	3.7	3.0	
80	*36,000*	13,700	**4.2**	3.7	
100	*64,000*	**21,200**	*5.0*	*4.6*	Not acceptable

Bold font—max permissible values (reference parameters for gG63).
[a]Not standardized by IEC, standardized only by Polish local regulation.

On the basis of Fig. 7.2 it is possible to evaluate let-through energy and cutoff current which occur in a circuit with a miniature circuit-breaker of B-type. Table 7.4 presents a comparison of these parameters for two values of short-circuit current: $I_k = 4$ kA and 6 kA.

After analyzing the parameters presented in Table 7.4, one may conclude that for a short-circuit current equal to $I_k = 6$ kA the backup protection of the RCCB marked ⊣▭⊢ 6000 is not possible by any of the analyzed circuit-breakers. It means that RCCB of a rated current $I_{nRCCB} = 63$ A can be backup protected by a gG63 fuse, but cannot be backup protected even by an MCB of B25, whose rated current is only $I_{nMCB} = 25$ A! In many cases, both reference values—let-through energy $= 21{,}200\ A^2s$ and cutoff

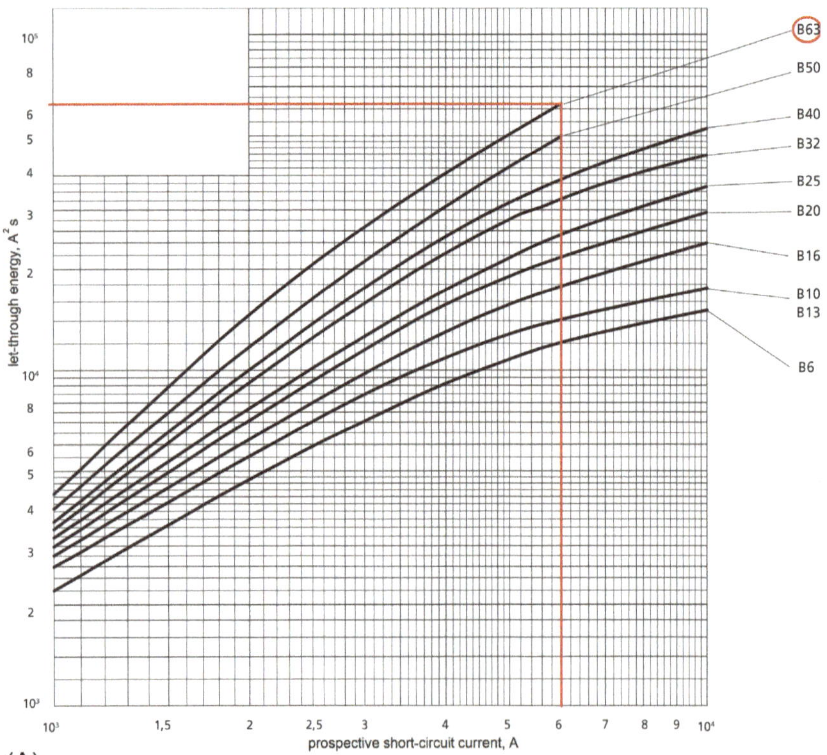

(A)

Fig. 7.2 Let-through energy (A) and cutoff current (B) of B-type miniature circuit-breakers [4], with marked example parameters for circuit-breakers B63 ($I_k = 6$ kA; ⇒ let-through energy ≈70,000 A^2s, cutoff current ≈4.6 kA).

(Continued)

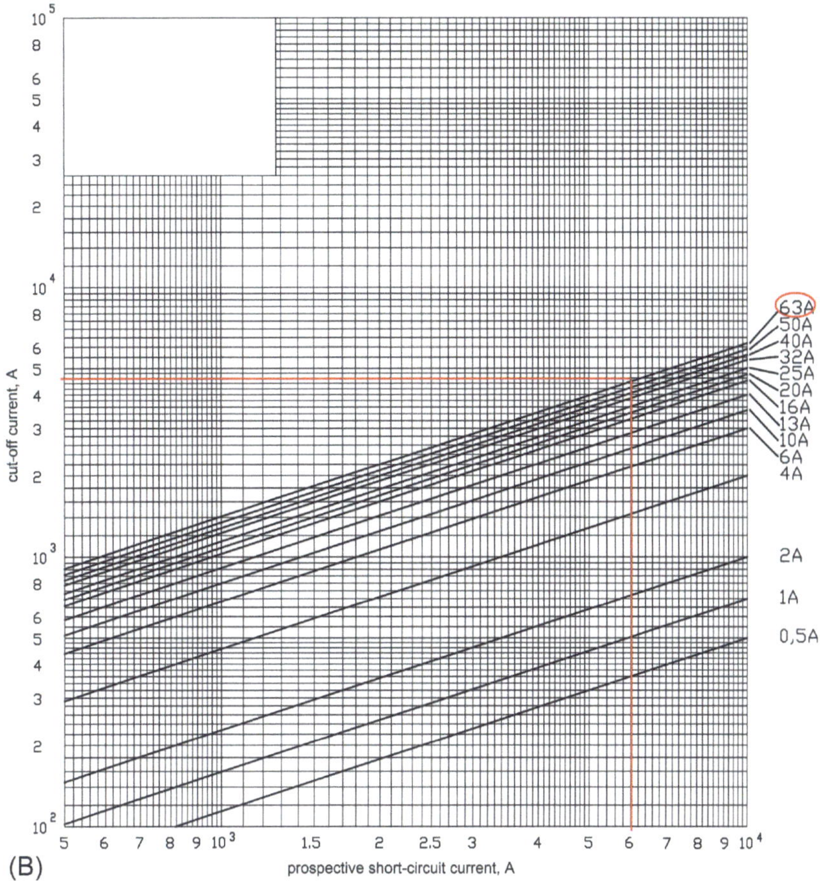

(B)

Fig. 7.2—Cont'd

current = 4.2 kA—are exceeded. Permissibility of backup protection of RCCB by MCB strictly depends on a value of short-circuit current. If a short-circuit current is equal to $I_k = 4$ kA, the use of circuit-breaker B25 is acceptable. Circuit-breakers of higher-rated current (B32 or more) cannot be used.

Bearing in mind the results of the earlier mentioned analysis, special attention has to be taken when RCCBs are considered to be backup protected by other overcurrent protective devices than provided by manufacturers in RCDs' datasheets.

Table 7.4 Characteristic parameters of B-type miniature circuit-breakers; prospective short-circuit current $I_k = 4\,kA$ and $6\,kA$; data according to [4].

	$I_k = 4\,kA$			$I_k = 6\,kA$		
Rated current of a circuit-breaker A	**Let-through energy $A^2 s$**	**Cutoff current kA**	**Backup protection**	**Let-through energy $A^2 s$**	**Cutoff current kA**	**Backup protection**
25	17,000	3.5	Acceptable	*24,000*	3.8	Not acceptable
32	*24,000*	3.6	Not acceptable	*32,000*	4.2	
40	*26,000*	3.7		*38,000*	*4.3*	
50	*32,000*	3.8		*50,000*	*4.4*	
63	*42,000*	3.9		*65,000*	*4.6*	

Max permissible values (reference parameters for gG63): let-through energy = 21,200 A^2 s; cutoff current = 4.2 kA.

References

[1] IEC 61008-1:2010, Residual current operated circuit-breakers without integral overcurrent protection for household and similar uses (RCCBs)—Part 1: General rules, 2010.
[2] IEC 60269-2:2013, Low-voltage fuses—Part 2: Supplementary requirements for fuses for use by authorized persons (fuses mainly for industrial application)—Examples of standardized systems of fuses A to K, 2013.
[3] ETI, Low-voltage fuse links, Technical data, ETI, 2019.
[4] ASTI, Miniature circuit-breakers, Technical data ASTI, ETI, 2017.

CHAPTER 8

Circuits with nonsinusoidal currents

8.1 General information

Selection of residual current devices has to be performed taking into account possible waveform shapes of the earth fault current, which may occur in the protected circuits. Detection of a residual current mainly relies on the proper transformation of this current by an inductive current transducer (iron core current transformer—Fig. 1.10B) being part of residual current devices. This transformed current affects the electromechanical relay and in the effect the residual current device opens the protected circuit. Therefore, in case of incorrect selection of residual current devices, they may not react at all to the residual current many times higher than their rated residual operating current [1–10]. Such risk of no reaction of the residual current device exists if its type is AC (detection of 50/60 Hz sinusoidal waveform only—see Table 3.4) and protects a rectifier circuit giving a relatively smooth direct residual current. The smooth direct current cannot be transformed properly by the iron core current transformer of AC-type RCDs. The next parts of this chapter present the behavior of residual current devices when nonsinusoidal residual currents flow.

8.2 Rectifiers

The widespread use of various types of devices containing rectifiers makes that unidirectional earth fault current is expected to be in industrial as well as domestic electrical installations. In the case of earth fault in a rectifier circuit, waveform shape depends on the type of a rectifier that is used in this circuit. From the point of view of RCDs, the most important is the pulsation of the residual current. The less pulsation of the residual current the more difficult is its detection. Table 8.1 presents the types of rectifiers and their corresponding earth fault (residual) currents.

In a half-wave rectifier and in a full-wave rectifier without smoothing supplied from L-N, residual currents are characterized by relatively high

Residual Current Devices
https://doi.org/10.1016/B978-0-323-89783-9.00003-5

Table 8.1 Residual current waveforms i_Δ in rectifier circuits; according to [11].

No.	Rectifier type	Simplified diagram	Residual current waveform
1	Half-wave rectifier	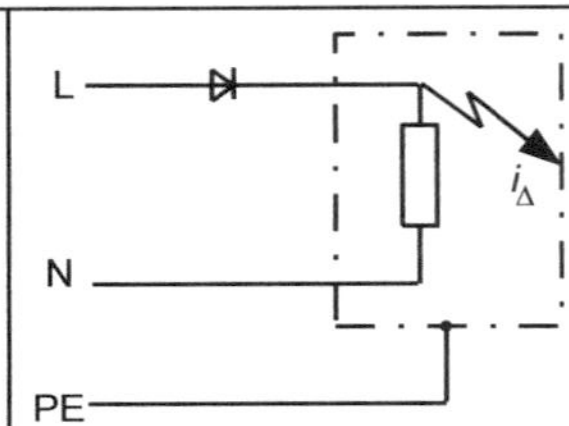	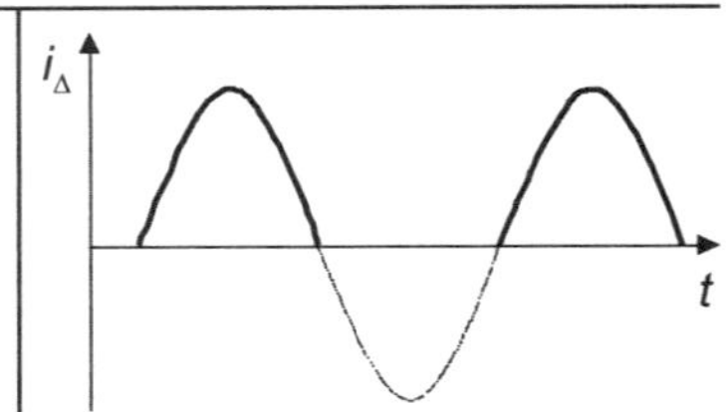
2	Full-wave rectifier supplied L-N, half controlled	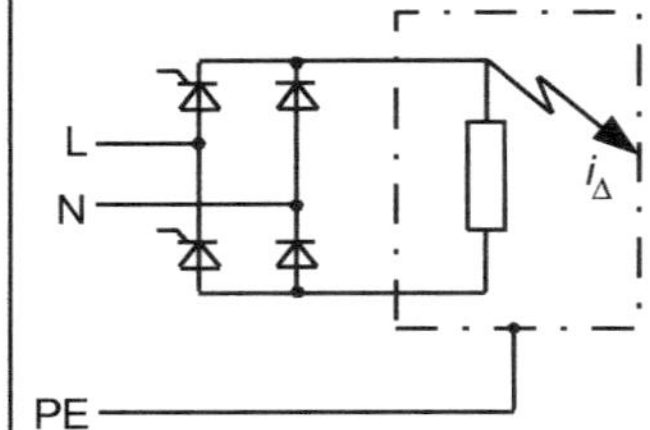	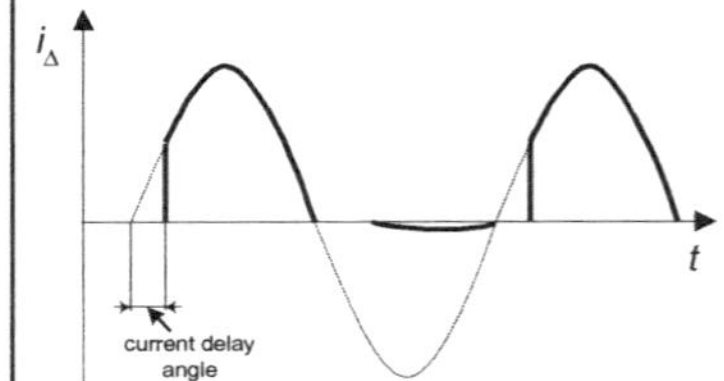
3	Half-wave rectifier with smoothing	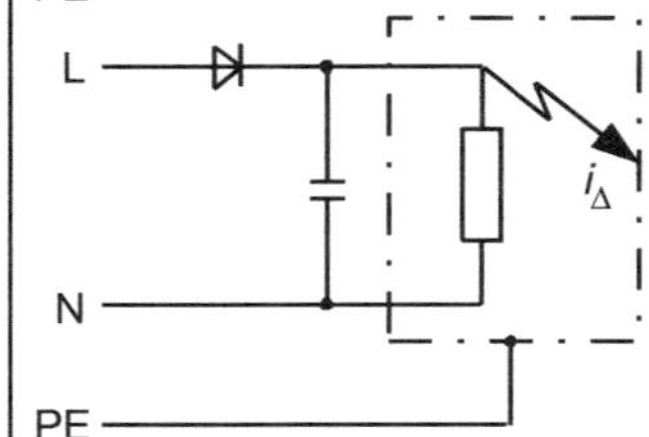	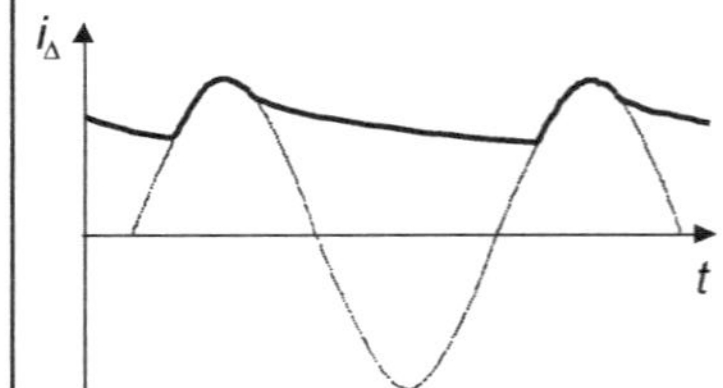
4	Two-pulse bridge supplied from two phases (L1-L2)	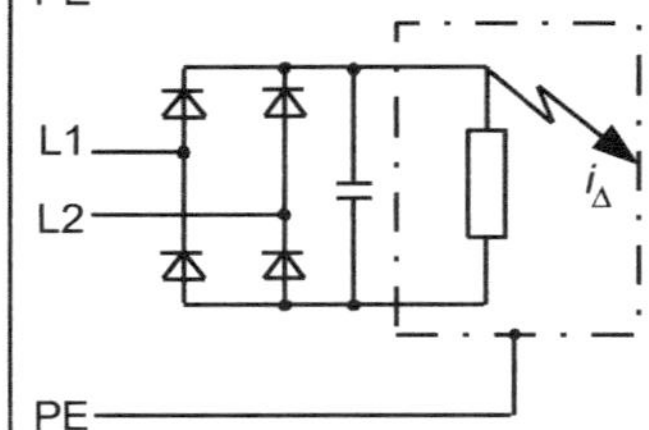	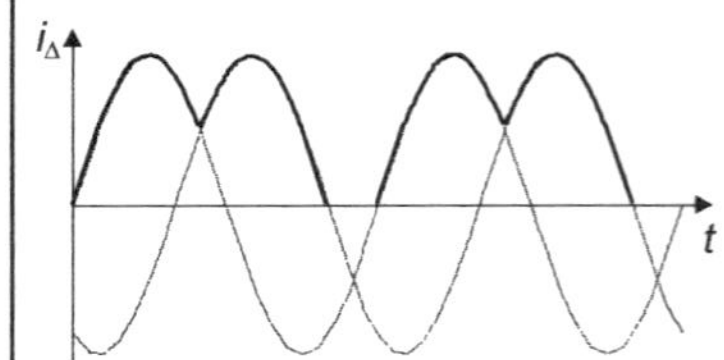

5	Six-pulse bridge	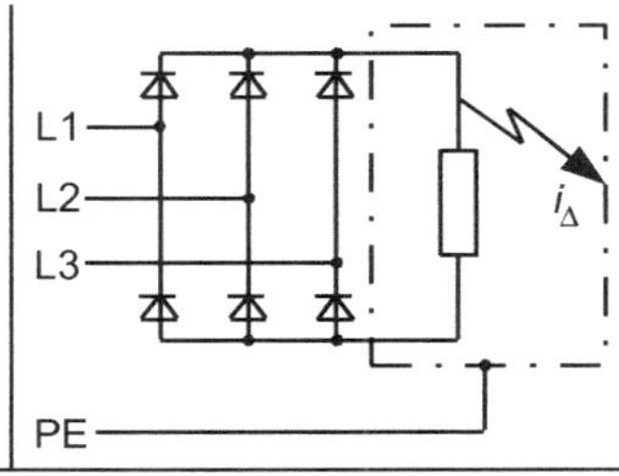	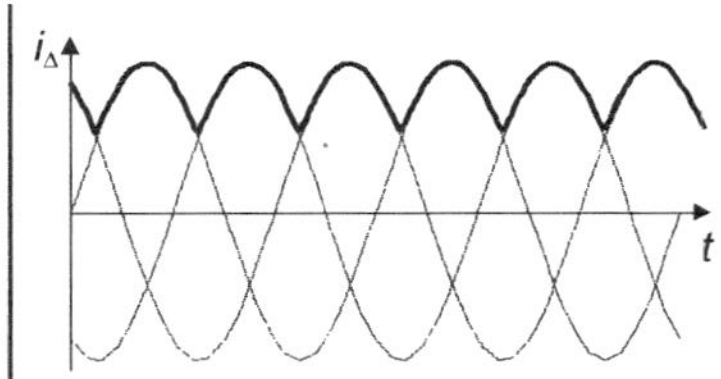

pulsation. It gives the possibility for detection of such residual currents by A-type and F-type RCDs because these RCDs contain current transformers with an iron core of the flat hysteresis loop (Fig. 2.5C). A flat hysteresis loop enables the production of sufficient variation of magnetic induction and magnetic flux in the core. When the rectifier includes a smoothing system or is three-phase supplied, residual currents are smooth and only B-type (or B+ type) RCDs are suitable.

The effect of residual current waveform pulsation on tripping of RCDs has been observed by the superimposition of smooth DC component on the sinusoidal waveform. In this laboratory test, the tripping current of A-type RCDs of $I_{\Delta n}=30\,mA$ was checked. The following values of DC component were consecutive superimposed on the sinusoidal waveform: 15, 30, 60, 90 and 150 mA. The higher value of DC component, the smoother the composite waveform. Fig. 8.1 depicts simulations of waveforms composed of the AC component equal to $I_{AC}=30\,mA$ (rms) and the aforementioned DC component I_{DC} of values from 15 mA to 150 mA. After forcing one of the aforementioned values of DC component, the sinusoidal current was increased in order to trip the tested RCD. Selected results of laboratory test are presented in Fig. 8.2. Tripping current in the vertical axis informs about rms value of the sinusoidal waveform being a part of the composite testing current.

For RCD no. 1, the tripping current (rms value of the sinusoidal component) does not exceed 30 mA when the DC component is equal to even 90 mA. This RCD is more resistant to DC component than RCD no. 2. For RCD no. 2, the DC component already equal to 60 mA gives tripping of the RCD for the sinusoidal component around 30 mA, which is the rated residual operating current of the tested RCD.

Very interesting is the residual current waveform generated by a two-pulse bridge supplied from two phases (Table 8.1, No. 4). This waveform has relatively high pulsation, however, A-type and F-type RCDs are not able to detect it. The half-wave presented in Fig. 8.3A enables for sufficient demagnetization of the iron core (during the dead time) before the next cycle starts. Such demagnetization ensures a relatively low level of remanence induction B_{rem} in the iron core characterized by the flat hysteresis loop (Fig. 2.5C). Unfortunately, in the case of the waveform depicted in Fig. 8.3B, the time for demagnetization of the iron core is too short and the level of magnetic induction B_{rem} before the next cycle is too high to generate sufficient variation of magnetic induction and magnetic flux.

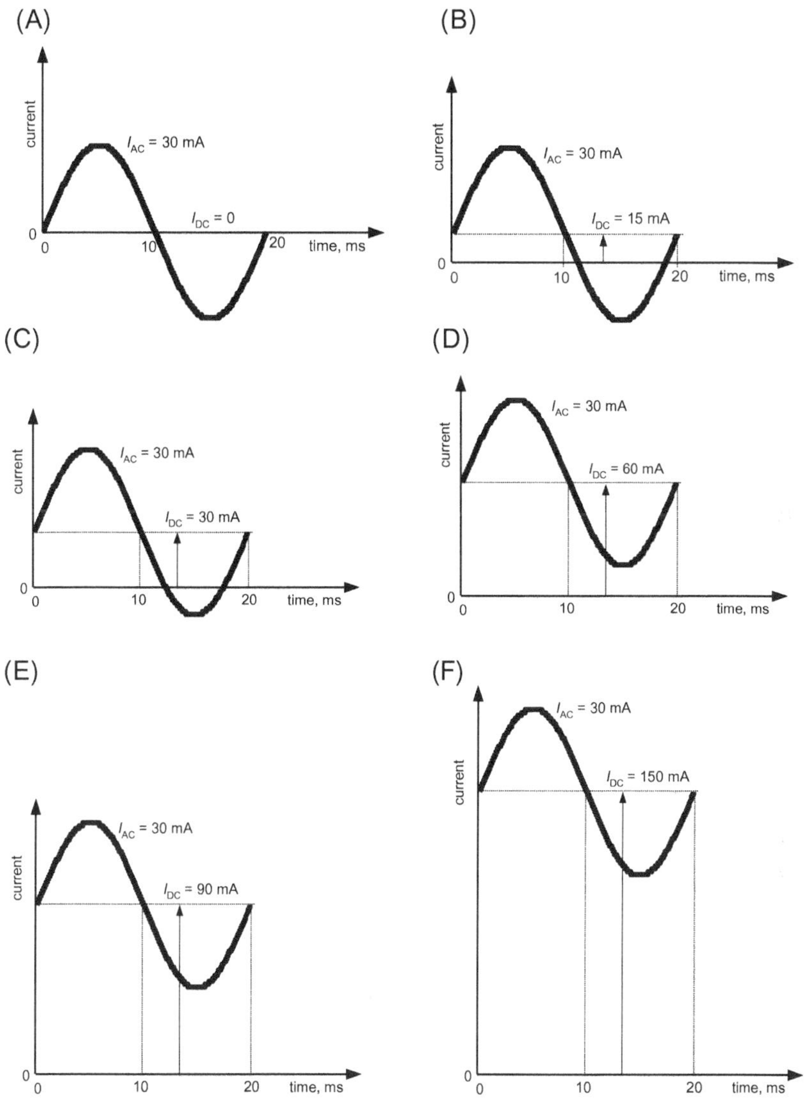

Fig. 8.1 Waveforms composed of sinusoidal AC component equal to 30 mA (I_{AC} = 30 mA) and the following values of DC component (I_{DC}): (A) 0 mA, (B) 15 mA, (C) 30 mA, (D) 60 mA, (E) 90 mA, (F) 150 mA.

Even more difficult to detect by an iron core current transformer is the waveform presented in Fig. 8.3C. This is a load current (not residual current) of a full-wave rectifier, but using this example it is very easy to show the effect of the dead time of unidirectional current waveform on tripping of RCDs.

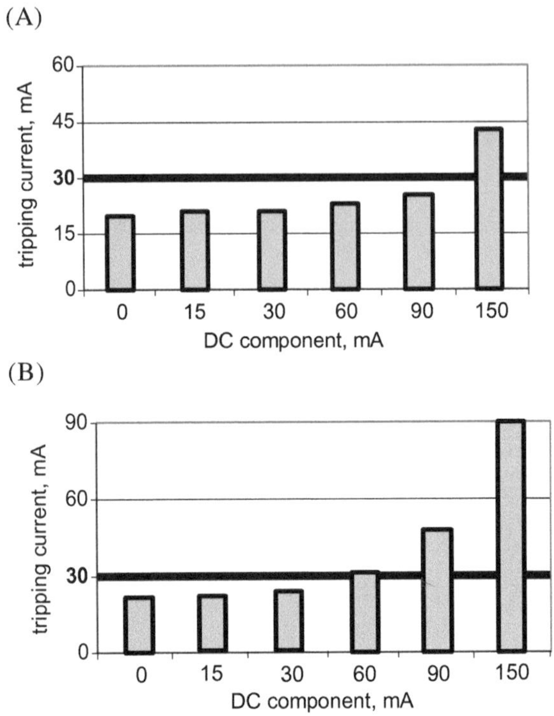

Fig. 8.2 Tripping current (rms value of the sinusoidal component) of A-type 30 mA RCDs for waveform composed of DC component + sinusoidal component: (A) RCD no. 1, (B) RCD no. 2.

In order to check the effect of waveform dead time (time for demagnetization) on the tripping of RCDs, a laboratory test has been performed. RCDs have been tested under the following current waveforms:

- sinusoidal,
- half-wave rectified,
- full-wave rectified.

For each waveform, the tripping current was checked. The residual current was increased slowly and the rms value of the tripping current was recorded. Results of the laboratory test are presented in Fig. 8.4. According to the standards [12,13], the real tripping current of A-type RCDs should be within the range:

- $(0.5–1.0)I_{\Delta n}$ for sinusoidal waveform,
- $(0.35–1.4)I_{\Delta n}$ for the half-wave rectified waveform.

When a sinusoidal residual current was forced, the real tripping current of both RCDs was within the permissible range. The same conclusion flows

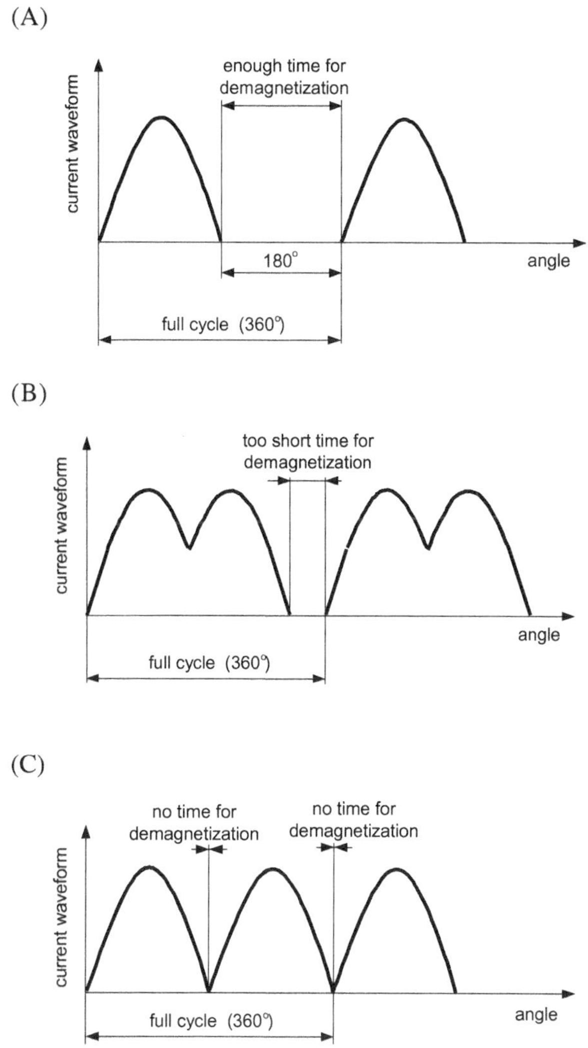

Fig. 8.3 Comparison of properties of various waveforms: (A) residual current in a circuit with half-wave rectifier, (B) residual current in a circuit with two-pulse bridge supplied from two phases, (C) load current in a circuit with a full-wave rectifier.

after checking this current under the half-wave rectified waveform. For the full-wave rectified waveform, the tripping current significantly exceeds even $1.4I_{\Delta n}$, despite the same pulsation as for the half-wave. It shows that the aforementioned dead time (time for demagnetization) is very important in the detection of unidirectional residual currents, not only the residual current pulsation.

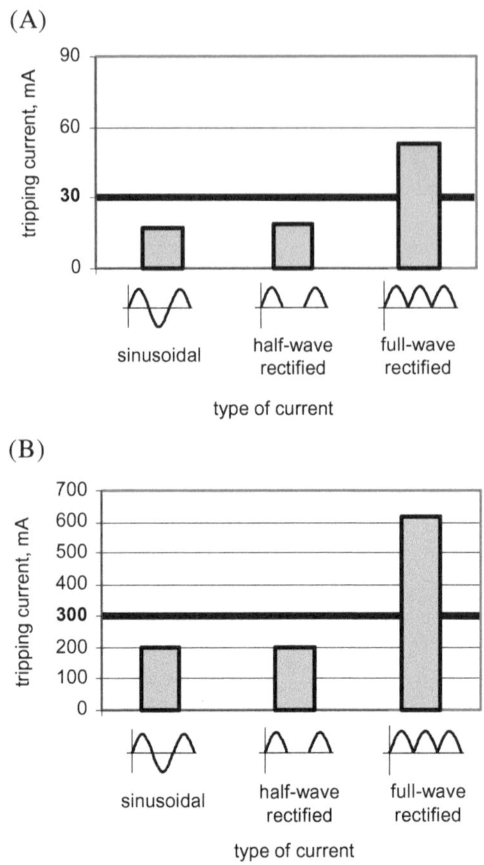

Fig. 8.4 Tripping current of A-type RCDs for various waveform shapes of laboratory residual currents: (A) RCD of $I_{\Delta n} = 30\,mA$, (B) RCD of $I_{\Delta n} = 300\,mA$.

Thus, for the detection of unidirectional waveforms characterized by relatively low pulsation as well as unidirectional waveforms characterized by very short dead time (time for demagnetization), only B-type and B+ type RCD are suitable.

When in a part of an electrical installation smooth DC earth fault current may flow, coordination between RCDs has to be maintained. Fig. 8.5 presents the requirements of standard EN 50178 [14] for this coordination. Circuit(s) of smooth DC current should be protected by B-type RCD and fed from the point upstream of the A-type RCDs. This is necessary for the proper operation of the A-type RCD in Fig. 8.5. If this requirement is not respected, the delayed A-type RCD will not detect a DC pulsating

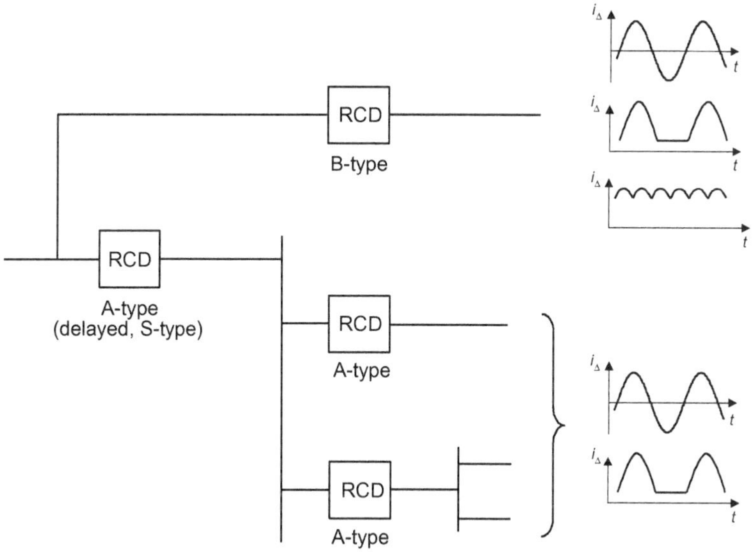

Fig. 8.5 Coordination between RCDs when in a part of the installation smooth DC earth fault current may flow; according to [14].

current in the main circuit, in case of the simultaneous superimposition of smooth DC component flowing due to an earth fault in circuit with B-type RCD.

8.3 Alternating residual currents with harmonics

Alternating residual currents with harmonics are mainly met in circuits with variable-speed drives, which comprise a frequency converter. The frequency converter for control of the motor speed is composed of a rectifier, an intermediate DC circuit, and an inverter (Fig. 8.6). In the case of an earth fault at the motor terminals, the earth fault current is distorted and comprises harmonics, whose order mainly depends on the PWM (Pulse Width Modulation) frequency. The earth fault current comprises a low-frequency component, which depends on the desired motor speed, a constant 150 Hz component as well as a component at the mentioned PWM frequency with its multiple, and also its interharmonics.

Figs. 8.7 and 8.8 present results of the computer simulations of the earth fault current waveform shapes and the current spectrum in case of a line-to-earth fault through the resistance 1 kΩ (conventional resistance of the human body) at the output terminals of the inverter when motor fundamental

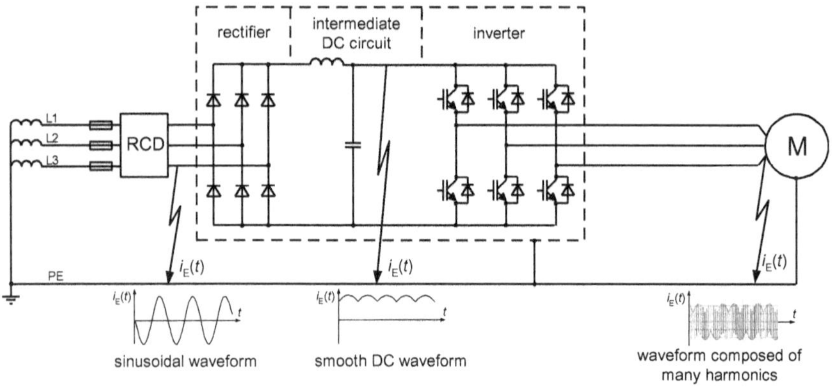

Fig. 8.6 Variable-speed drive circuit and possible waveform shape of the earth fault current $i_E(t)$.

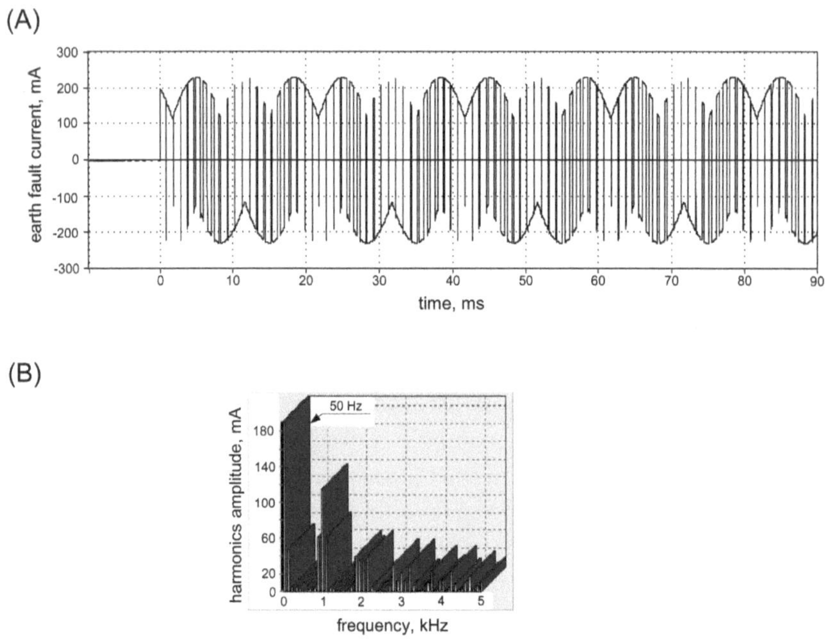

Fig. 8.7 Earth fault currents waveform (A) and its spectrum (B) for motor frequency 50 Hz and PWM frequency 1 kHz.

supplying frequency is equal to 50 Hz (rated motor speed). The earth fault current waveform is mainly composed of the 50 Hz component and the PWM component (1 kHz—Fig. 8.7B; 3 kHz—Fig. 8.8B). There are also other high-frequency components, which come from PWM frequency.

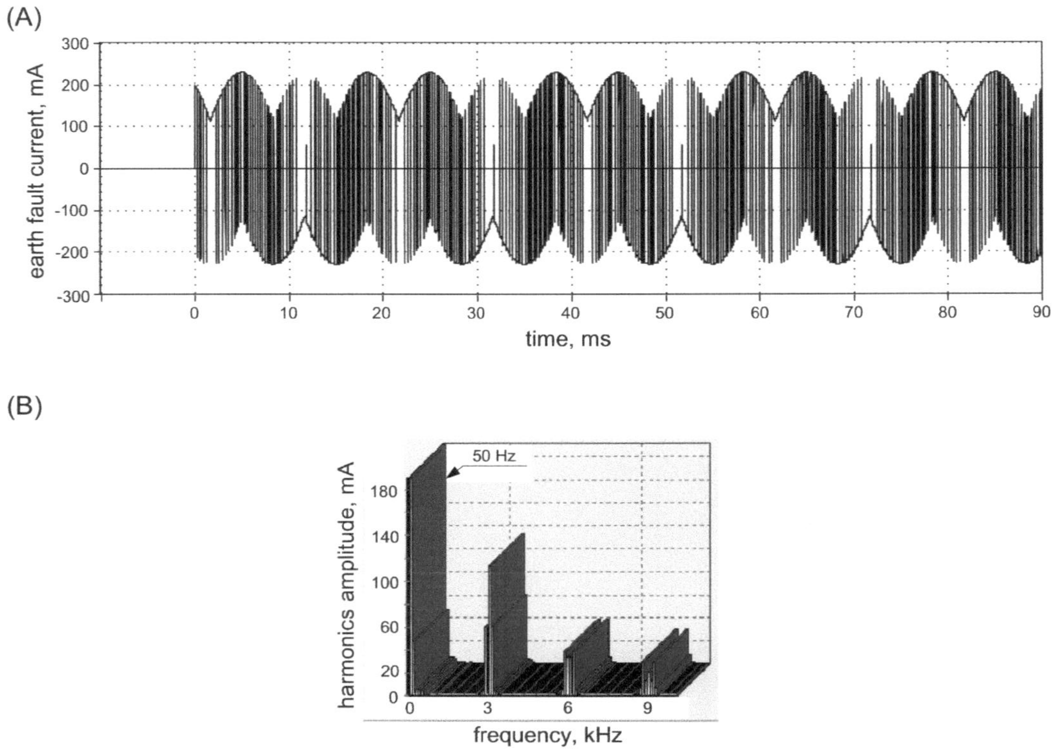

Fig. 8.8 Earth fault currents waveform (A) and its spectrum (B) for motor frequency 50 Hz and PWM frequency 3 kHz.

It should be underlined that the amplitude of the component of motor frequency and the PWM component change as a function of the motor speed and its operating frequency. In order to show this dependency, a laboratory test in a variable speed-drive circuit has been performed. In this circuit, the output terminals of the inverter were short-circuited to the earth via resistance 1 kΩ. The short-circuits were arranged for various values of the motor operating frequency, especially for:

- 50 Hz—rated motor speed,
- 25 Hz—half-rated motor speed,
- 5 Hz—very low motor speed.

Oscillograms of the earth fault current waveforms for the aforementioned frequencies are presented in Figs. 8.9–8.11.

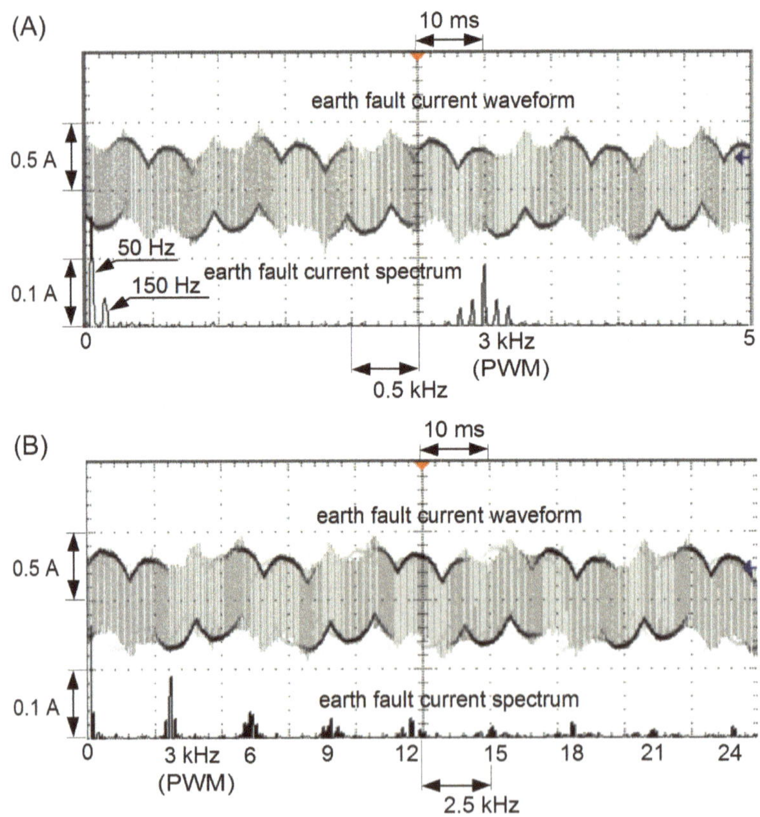

Fig. 8.9 Oscillograms of earth fault current waveforms and their spectrum for motor frequency 50 Hz and PWM frequency 3 kHz. The current spectrum within the range: (A) 0–5 kHz, (B) 0–24 kHz.

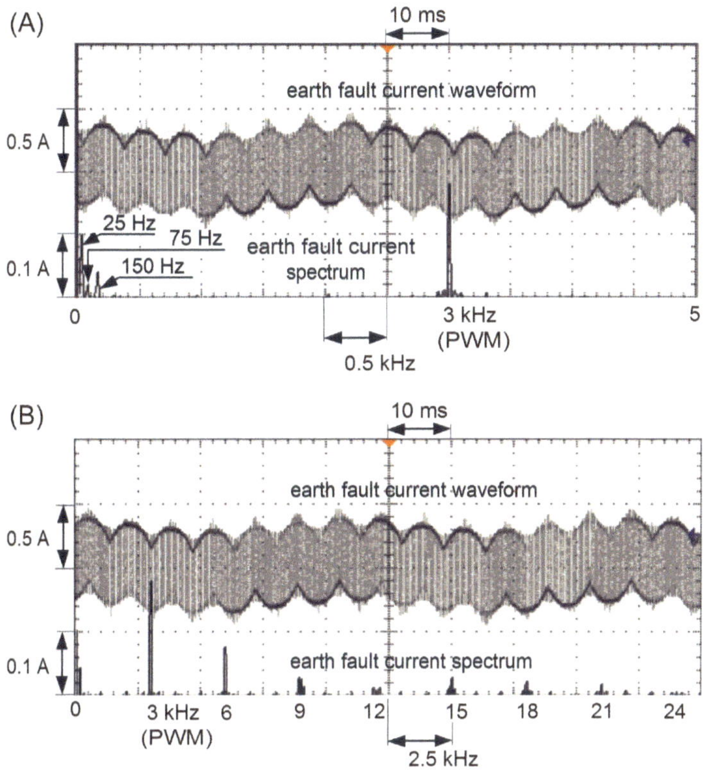

Fig. 8.10 Oscillograms of earth fault current waveforms and their spectrum for motor frequency 25 Hz and PWM frequency 3 kHz. The current spectrum within the range: (A) 0–5 kHz, (B) 0–24 kHz.

The earth fault current comprises a low-frequency component which depends on the desired motor speed (50 Hz—Fig. 8.9; 25 Hz—Fig. 8.10, and 5 Hz—Fig. 8.11), a constant 150 Hz component as well as a component at the PWM frequency (here 3 kHz), its multiple and also its interharmonics. The amplitude of the current of motor frequency ("operating frequency") and the PWM current change as a function of the motor speed and frequency. For the 50 Hz operating frequency the amplitude of the 50 Hz component exceeds the amplitude of the PWM component (Fig. 8.9), but for a low motor frequency, the participation of the above-mentioned components varies in the opposite ways [1,15,16]. For very low motor frequency the amplitude of the PWM component significantly exceeds the amplitude of current of motor frequency and the 150 Hz component (Fig. 8.11). Fig. 8.12 presents a variation of the earth fault current components as a function of motor operating frequency.

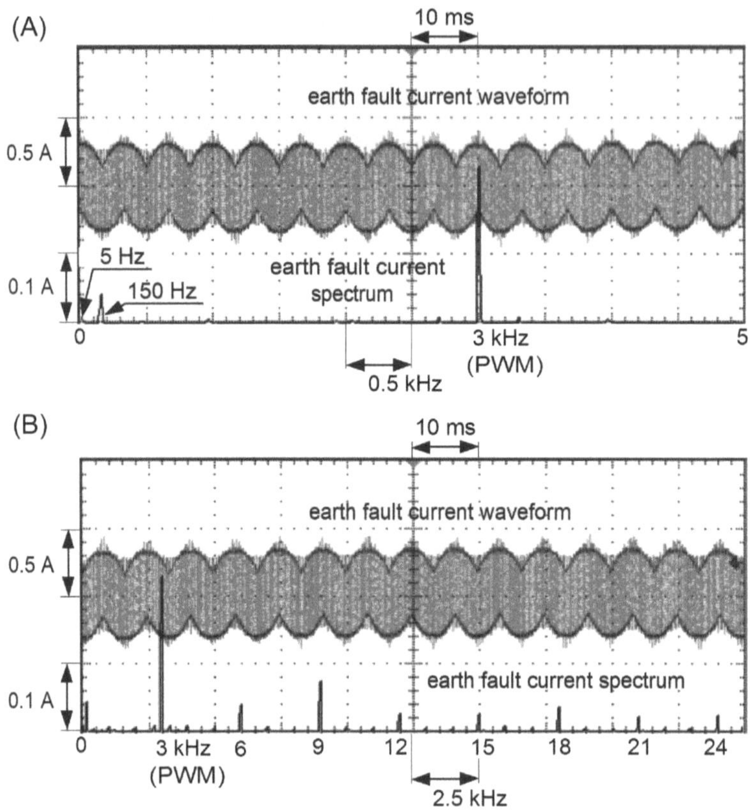

Fig. 8.11 Oscillograms of earth fault current waveforms and their spectrum for motor frequency 5 Hz and PWM frequency 3 kHz. The current spectrum within the range: (A) 0–5 kHz, (B) 0–24 kHz.

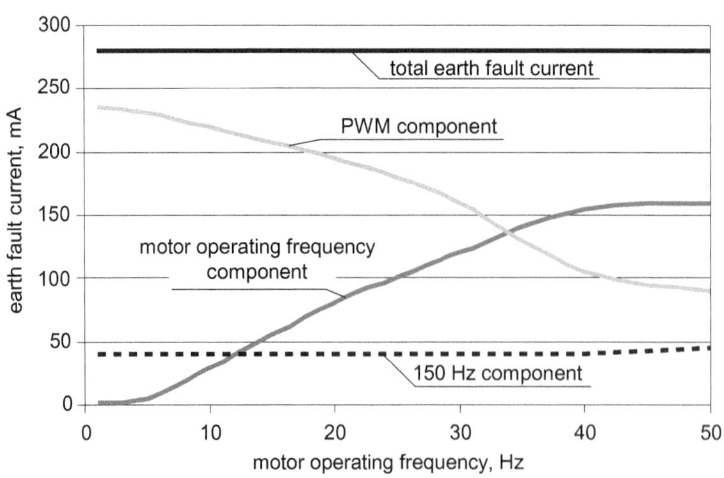

Fig. 8.12 Variation of earth fault current components as a function of motor operating frequency in a variable-speed drive circuit.

As one can see, current waveforms with different spectra can be expected in circuits with power electronic converters. Tripping current of the RCDs depends on the spectrum of the earth fault current—in particular, the order of harmonics is important and in some cases their phase angle.

The distorted secondary current i_s (I_s—see Fig. 2.1), which flows in the coil of the electromechanical relay, can be described by the following equation:

$$i_s(t) = \sqrt{2} \cdot I_s \cdot [\sin(\omega t + \alpha_{h1}) + K_{h3} \sin(3\omega t + \alpha_{h3}) + K_{h5} \sin(5\omega t + \alpha_{h5}) + \ldots + K_{hn} \sin(n\omega t + \alpha_{hn})] \quad (8.1)$$

where

K_{h3}, K_{h5}, ... K_{hn}—coefficient amplitude-to-fundamental of 3rd, 5th, ..., nth harmonic,

α_{h3} α_{h5}, ... α_{hn}—phase angle of 3rd, 5th, ..., nth harmonic.

The following theoretical analysis does not take into account the fact that the secondary circuit winding and electromechanical relay impedance vary with the frequency and harmonics. The impact of the phase angle of a particular harmonic on the electromechanical relay operation is only considered.

The effect of a single harmonic on the electromechanical relay operation is considered, as a representative example, for:

- relatively low-order harmonic (3rd),
- relatively high-order harmonic (49th),

and the following amplitudes of harmonic-to-fundamental:

- the 3rd harmonic: $K_{h3} = 0.1$, $K_{h3} = 0.25$ and $K_{h3} = 0.5$,
- the 49th harmonic: $K_{h49} = 0.1$, $K_{h49} = 0.25$ and $K_{h49} = 0.5$.

The harmonics can have the same phase angle as the fundamental $\alpha_{h3} = 0°$ or the phase angle can be 180° ($\alpha_{h3} = 180°$).

The analysis of the electromechanical relay operation is presented in a graphical way, as shown in Figs. 8.13–8.18. Symbols in Figs. 8.13–8.18 indicate:

F_1—a resultant magnetic force for sinusoidal wave,

F_{1+3h}—a resultant magnetic force for fundamental and 3rd harmonic,

F_{1+49h}—a resultant magnetic force for fundamental and 49th harmonic,

$F_{N\text{-}S}$—magnetic force derived from the permanent magnet,

F_{sp}—equivalent magnetic force derived from the spring,

r.l.—reference level/line; its crossing indicates the RCD tripping,

P_{tr}—point, at which the RCD tripping occurs.

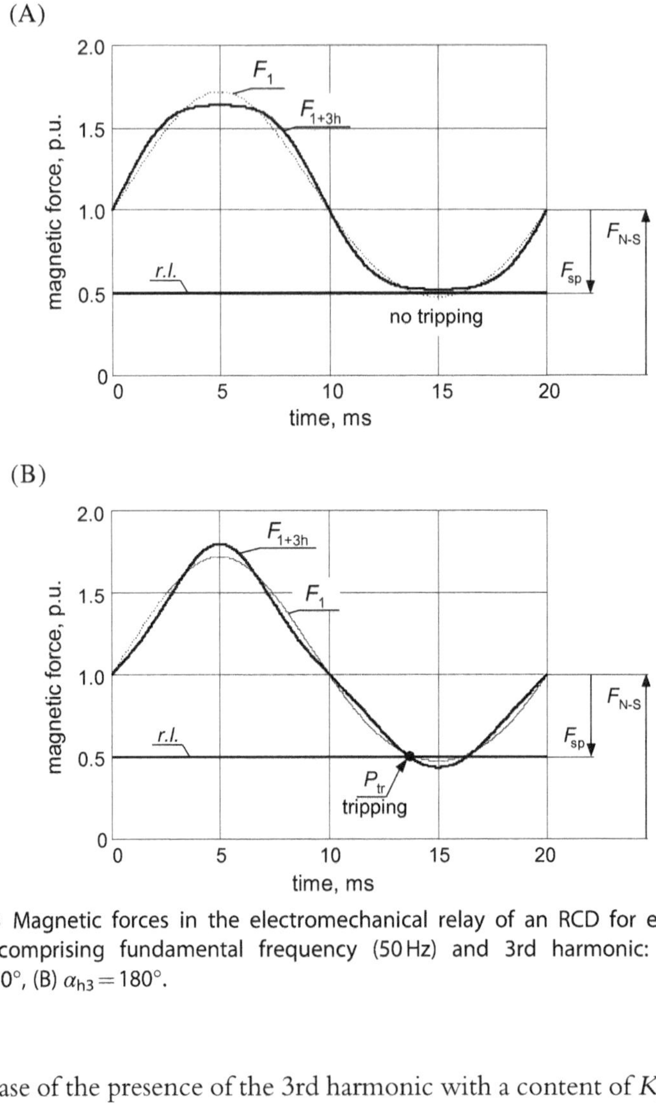

Fig. 8.13 Magnetic forces in the electromechanical relay of an RCD for earth fault current comprising fundamental frequency (50 Hz) and 3rd harmonic: $K_{h3}=0.1$; (A) $\alpha_{h3}=0°$, (B) $\alpha_{h3}=180°$.

In the case of the presence of the 3rd harmonic with a content of $K_{h3}=0.1$, its phase angle influences the electromechanical relay as well as the RCD tripping. The 3rd harmonic may decrease or increase the tripping point of the RCD. There is no tripping point P_{tr} in Fig. 8.13A because the trace representing magnetic force F_{1+3h} line does not cross the reference trace/level *r.l.* representing resultant force coming from the interaction "permanent magnet—spring" ($F_{N\text{-}S}-F_{sp}$). In order to cross that trace, the earth fault (residual) current must have a higher value or other phase angles (e.g., $\alpha_{h3}=180°$—see Fig. 8.13B). The waveform shape presented

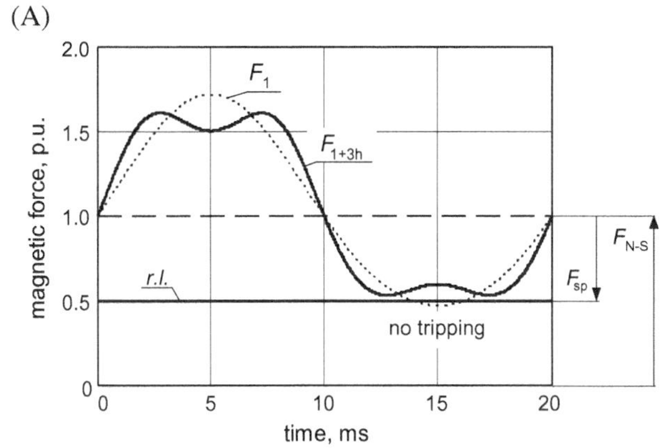

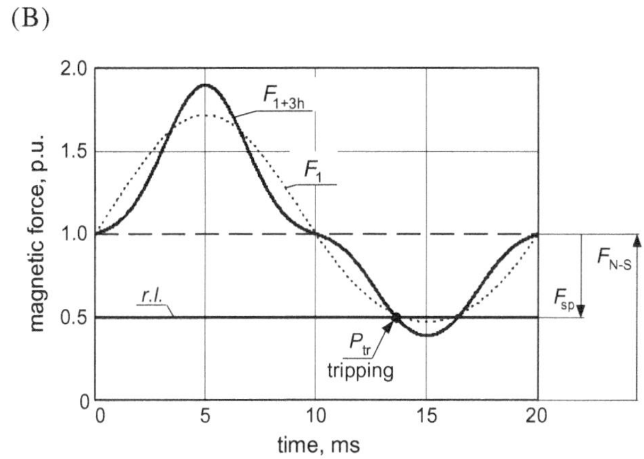

Fig. 8.14 Magnetic forces in the electromechanical relay of an RCD for earth fault current comprising fundamental frequency (50 Hz) and 3rd harmonic: $K_{h3}=0.25$; (A) $\alpha_{h3}=0°$, (B) $\alpha_{h3}=180°$.

in Fig. 8.13B increases the sensitivity of the RCD. In such a case, the RCD trip out for the rms earth fault current value which is lower than for the pure sinusoidal current.

The effect of the phase angle of the harmonic is even more pronounced in the case of the amplitude of harmonic-to-fundamental equal to $K_{h3}=0.25$ (Fig. 8.14). For the phase angle $\alpha_{h3}=0°$ (Fig. 8.14A), the distorted waveform F_{1+3h} has an increased distance to the reference level *r.l.* (the crossing of which means tripping of the RCD) compared to the case from Fig. 8.13A. The phase angle $\alpha_{h3}=180°$ (Fig. 8.14B), gives a significant excess of the reference level, so the conditions for the RCD tripping are favorable.

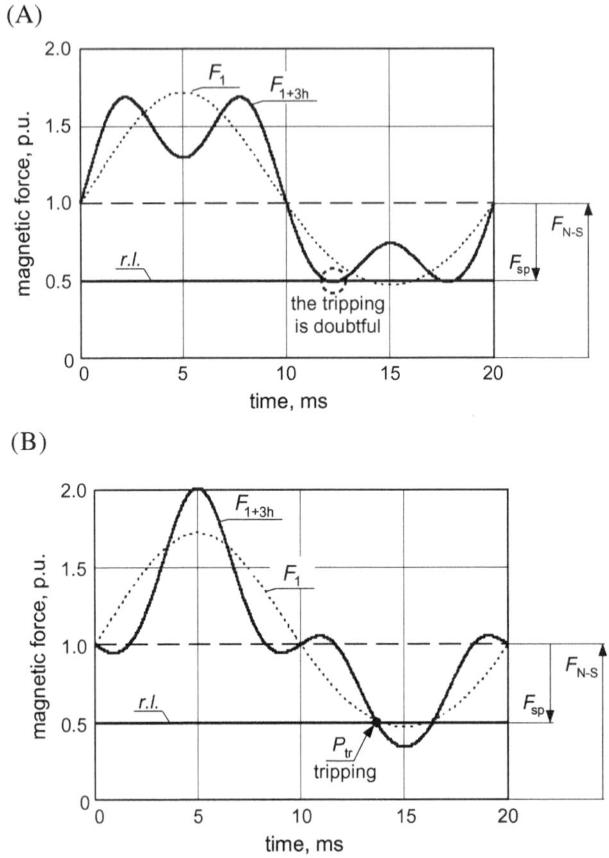

Fig. 8.15 Magnetic forces in the electromechanical relay of an RCD for earth fault current comprising fundamental frequency (50 Hz) and 3rd harmonic: $K_{h3}=0.5$; (A) $\alpha_{h3}=0°$, (B) $\alpha_{h3}=180°$.

A similar conclusion flows from the results of the analysis presented in Fig. 8.15, where the amplitude of harmonic-to-fundamental equal to $K_{h3}=0.5$ is considered. For the phase angle of the 3rd harmonic equal to 0° (Fig. 8.15A) the tripping of the RCD is not guaranteed, because the waveform F_{1+3h} only touches the reference line *r.l.* reflecting the tripping level. In the case of the phase angle $\alpha_{h3}=180°$ (Fig. 8.15B), the tripping occurs.

If the earth fault current contains harmonics of a considerably higher order, for example, 49th, the influence of the phase angle of that harmonic on RCD operation is negligible. Fig. 8.16 presents the analysis results when the amplitude of harmonic-to-fundamental is equal to $K_{h49}=0.1$.

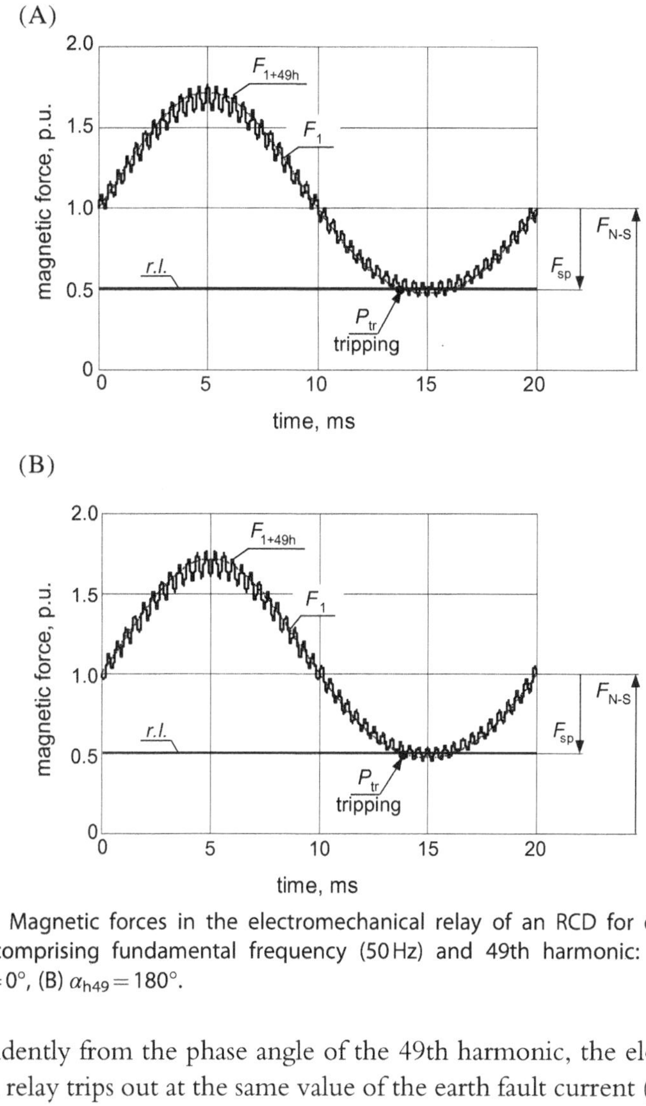

Fig. 8.16 Magnetic forces in the electromechanical relay of an RCD for earth fault current comprising fundamental frequency (50 Hz) and 49th harmonic: $K_{h49}=0.1$; (A) $\alpha_{h49}=0°$, (B) $\alpha_{h49}=180°$.

Independently from the phase angle of the 49th harmonic, the electromechanical relay trips out at the same value of the earth fault current (compare Fig. 8.16A vs. Fig. 8.16B).

Similar conclusions can be drawn on the basis of the results presented in Figs. 8.17 and 8.18. Regardless of whether the content of the 49th harmonic is $K_{h49}=0.25$ or $K_{h49}=0.5$, the tripping threshold of the RCD is at the same level. The phase angle of the harmonic is not important.

In order to confirm the aforementioned theoretical analysis, laboratory tests have been performed. During the first laboratory test, only a single harmonic was superimposed on a sinusoidal 50 Hz waveform. The test was

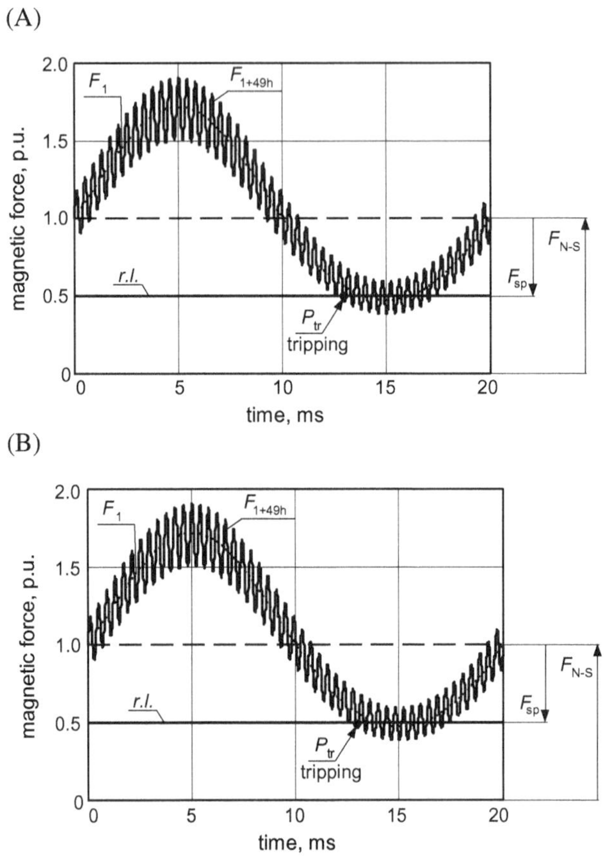

Fig. 8.17 Magnetic forces in the electromechanical relay of an RCD for earth fault current comprising fundamental frequency (50 Hz) and 49th harmonic: $K_{h49}=0.25$; (A) $\alpha_{h49}=0°$, (B) $\alpha_{h49}=180°$.

carried out for several orders of harmonics, mainly for: 3rd and 49th. The rms value of earth fault current was increased and the real tripping current was recorded. The phase angle of harmonic of the earth fault current varied with 45° step. Participation of the particular harmonic was the following: 0.10; 0.25; 0.50; 1.00, 2.00; (see in Figs. 8.19 and 8.20: 10%, 25%, 50%, 100%, 200%). In addition, to evaluate reference tripping level, tripping current was verified for pure sinusoidal current—this trace is marked "sin 50 Hz". Figs. 8.19 and 8.20 present the selected results of this test.

Results of the laboratory test are convergent with the theoretical analysis. If the earth fault current comprises low-order harmonics (here 3rd), the real tripping current may be higher or lower than for the pure sinusoidal

(A)

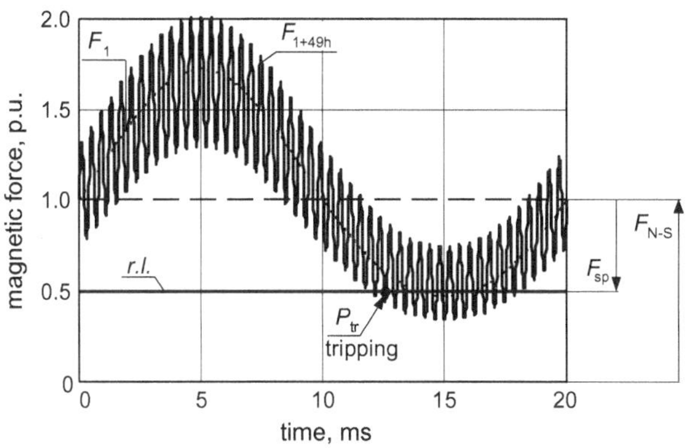

(B)

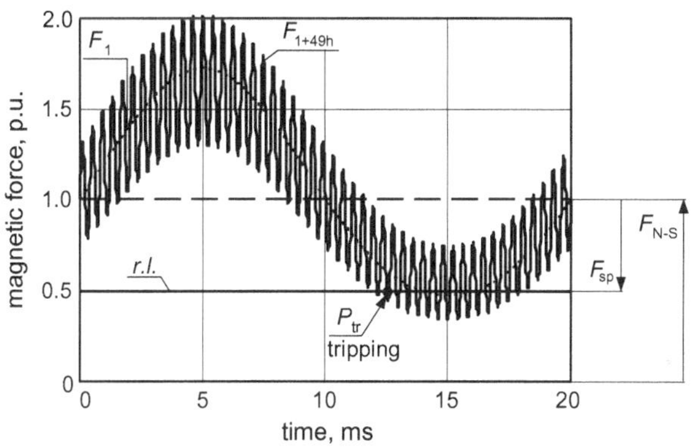

Fig. 8.18 Magnetic forces in the electromechanical relay of an RCD for earth fault current comprising fundamental frequency (50 Hz) and 49th harmonic: $K_{h49}=0.5$; (A) $\alpha_{h49}=0°$, (B) $\alpha_{h49}=180°$.

current—it depends on the harmonic phase angle. For a high-order harmonic (here 49th) tripping current does not depend on the harmonic phase angle.

However, high participation of a high-order harmonic makes the real tripping current of RCD very high with reference to the pure sinusoidal current. In the case of the tested AC-type RCD (Fig. 8.19B), the tripping current increases twice when 100% participation of the 49th harmonic occurs, but twelve times when 200% participation of the 49th harmonic occurs. It is very risky, and effective protection against electric shock may not be achieved.

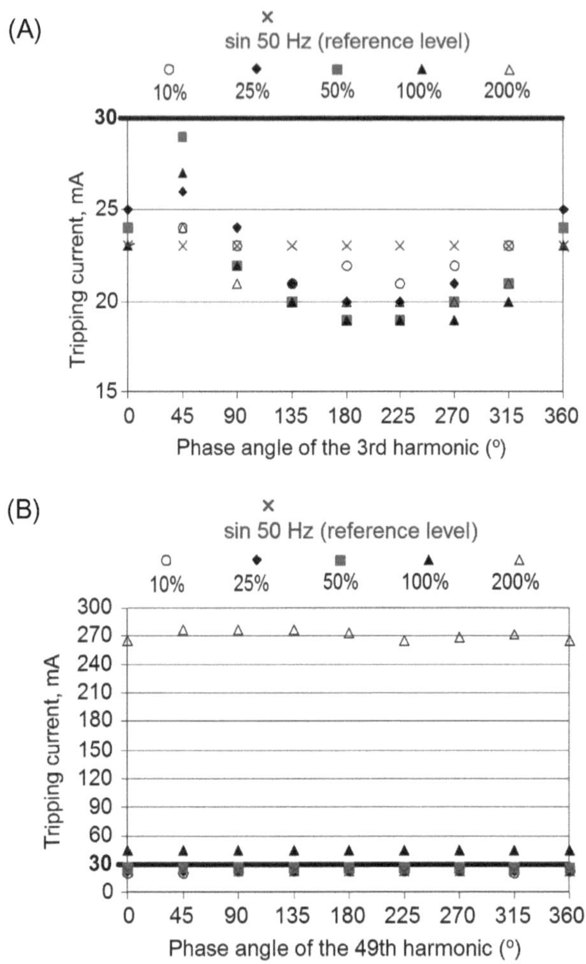

Fig. 8.19 Tripping current of the selected RCD (30 mA, AC-type) as a function of the phase angle of a harmonic. Earth fault current is composed of the following components: (A) fundamental +3rd harmonic, (B) fundamental +49th harmonic. Content of a harmonic (percentage of harmonic-to-fundamental): 10%, 25%, 50%, 100%, 200%.

The sources of the negative impact of the high-order harmonic are:

- the current transformer behavior,
- variation of the impedance of the secondary circuit winding as well as the electromechanical relay of the RCD.

Laboratory test results show that the hysteresis loop of some types of current transformers can be significantly wider for a higher frequency (Fig. 8.21). It is a negative effect because to achieve the same level of induction in the current transformer core for a higher frequency, a higher value of earth fault current is necessary. Moreover, the higher the earth fault current frequency

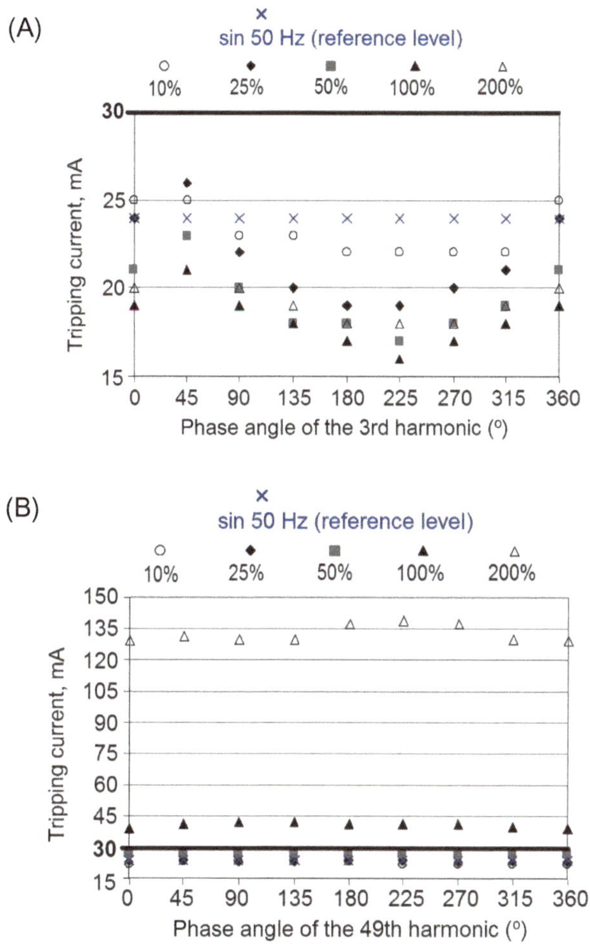

Fig. 8.20 Tripping current of the selected RCD (30 mA, A-type) as a function of the phase angle of a harmonic. Earth fault current is composed of the following components: (A) fundamental +3rd harmonic, (B) fundamental +49th harmonic. Content of a harmonic (percentage of harmonic-to-fundamental): 10%, 25%, 50%, 100%, 200%.

the higher is the secondary circuit winding and the electromechanical relay impedance. Fig. 8.22 presents results of the frequency-impedance test of the example electromechanical relay dismounted from a 30 mA RCD (A-type).

It is seen that the impedance of the relay is strongly dependent on the frequency of the current flowing through the relay winding (coil). The change of the frequency from 50 Hz to 1000 Hz results in the change of the impedance around 13 times. Such an effect of the frequency on the key elements of the RCD (current transformer and electromechanical relay) may result in its higher tripping threshold. Figs. 8.23 and 8.24 present the tripping threshold of the four RCDs in the frequency domain. The test

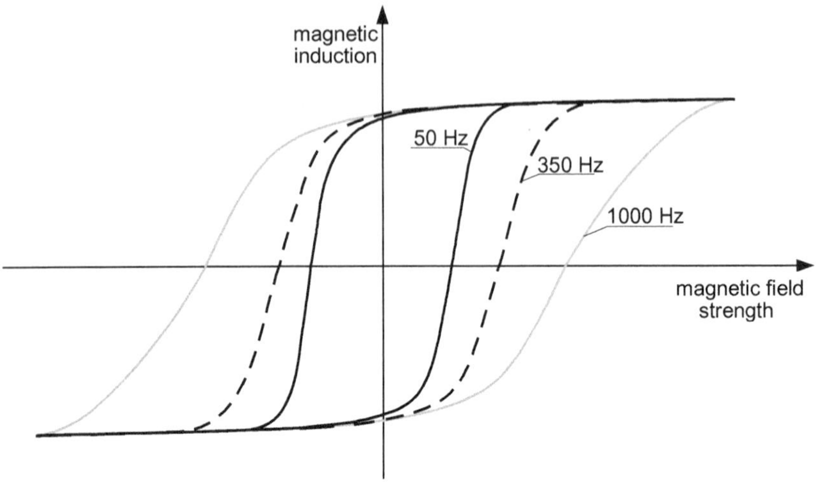

Fig. 8.21 Diagram of changes in the shape of the hysteresis loop of the selected current transformer (dismounted from a 300 mA AC-type RCD) for the following frequencies: 50 Hz, 350 Hz and 1000 Hz.

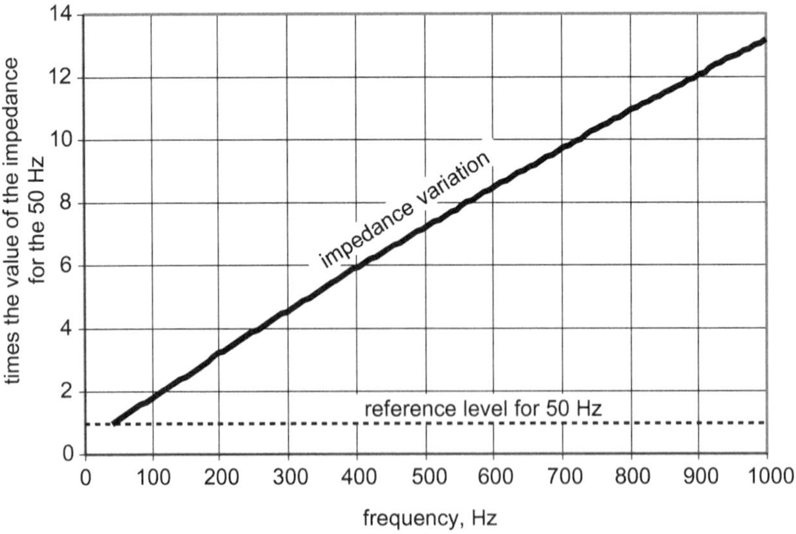

Fig. 8.22 Variation of the impedance of the selected electromechanical relay dismounted from a 30 mA A-type RCD.

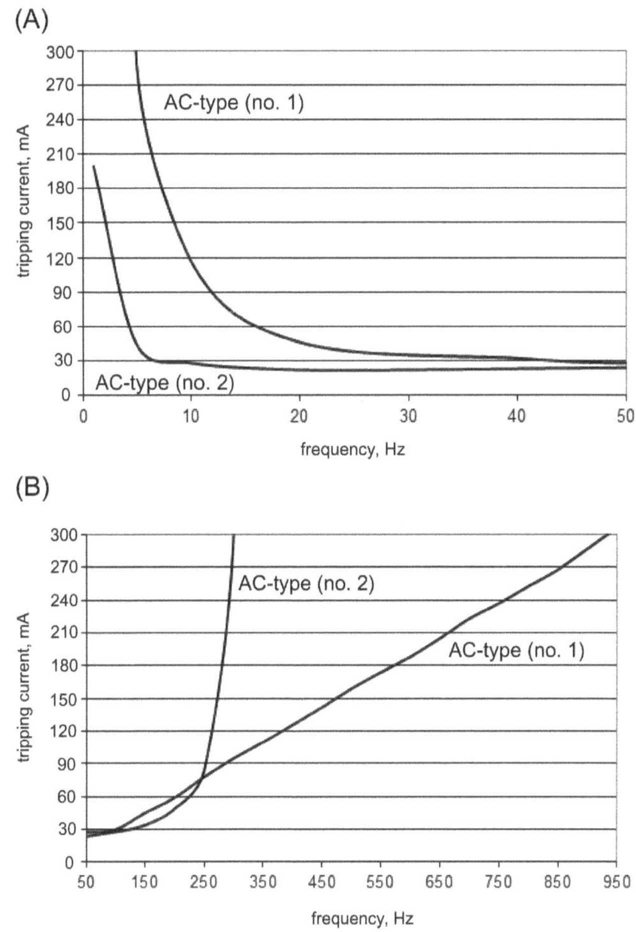

Fig. 8.23 Tripping current of the selected AC-type RCDs ($I_{\Delta n}=30\,\text{mA}$) as a function of frequency; the frequency range: (A) 1–50 Hz, (B) 50–950 Hz.

was performed for frequencies lower (Figs. 8.23A and 8.24A) and higher (Figs. 8.23B and 8.24B) than 50 Hz.

Thus, the tripping current of RCDs may be higher than the one for the rated frequency residual current, if the high content of a high-order harmonic is present in an earth fault (residual) current.

The tripping current of RCDs has also been checked in a real circuit with a variable speed drive (the second laboratory test with harmonics). In such a circuit, earth fault (residual) current is composed of many harmonics, as it is presented in Figs. 8.7–8.12. Moreover, the content of harmonics depends on the actual motor speed. Fig. 8.25 presents the tripping current of two selected RCDs for various motor speeds and PWM frequency equal to 3 kHz. One phase output terminal of the inverter was connected to the earth

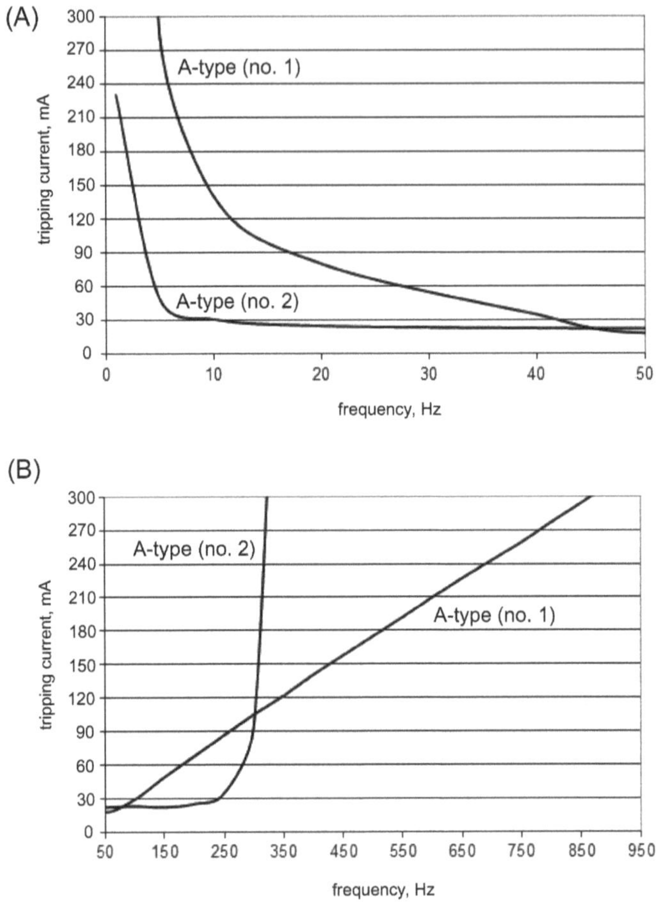

Fig. 8.24 Tripping current of the selected A-type RCDs ($I_{\Delta n}$ = 30 mA) as a function of frequency; the frequency range: (A) 1–50 Hz, (B) 50–950 Hz.

via variable resistance, and the tripping current of the tested RCD was recorded. RCDs sensitivity becomes worse (higher tripping current) as motor speed decreases. When motor speed is very low (5 Hz and 1 Hz), RCD2 did not trip at all. This is due to high-frequency current components, which dominate in earth current waveforms in the case of very low motor speed (Figs. 8.11 and 8.12).

Circuits characterized by earth fault current composed of many harmonics, especially very high-order harmonics, should be protected by RCDs of known tripping characteristics as a function of residual current frequency. Fig. 8.26 presents examples of such tripping characteristics.

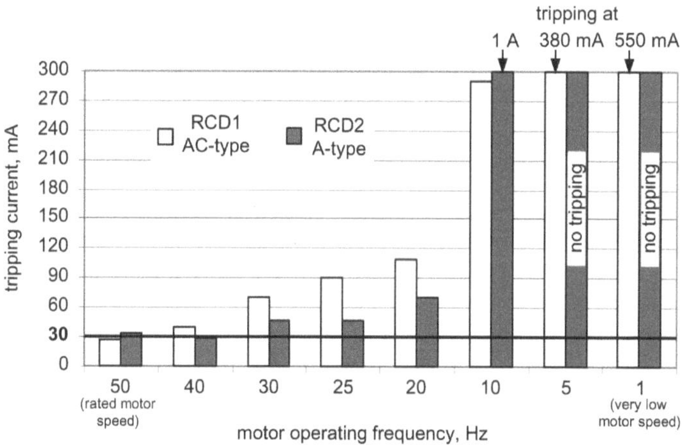

Fig. 8.25 Tripping current of the selected RCDs (30 mA, AC-type and A-type) as a function of motor operating frequency of the variable-speed drive.

8.4 Alternating residual currents characterized by current delay angle (symmetrical current control)

Earth fault current distortion can also be characterized by the current delay angle. Converters with current delay angle (symmetrical current control) are used for control of current-using equipment power and can be met in various types of installations, including domestic installations. A simplified circuit with such a type of converter, as well as an earth fault current waveform shape are presented in Fig. 8.27.

The higher the value of the current delay angle, the more distorted the current. However, from the point of view of residual current devices tripping, the fact that the current waveform is bidirectional is very important. Bidirectional current coming from the 50/60 Hz waveform, in spite of the high distortion (high delay angle value), can be easy detected even by the simplest type of RCDs—AC-type RCD, because sufficient variation of magnetic induction in the iron core of the current transformer occurs [19].

Fig. 8.28 presents induced secondary voltage in a selected current transformer of the RCD. After comparison of Fig. 8.28A with Fig. 8.28B it may be concluded that primary (earth fault) current with delay angle equal to 90° gives a higher peak value of the induced secondary voltage than the sinusoidal primary current. The tripping current of the electromechanical relay of RCDs depends on the peak value of current flowing through this relay. Thus, it is suspected that the tripping threshold of RCDs for high current delay angle can be lower than for the sinusoidal current.

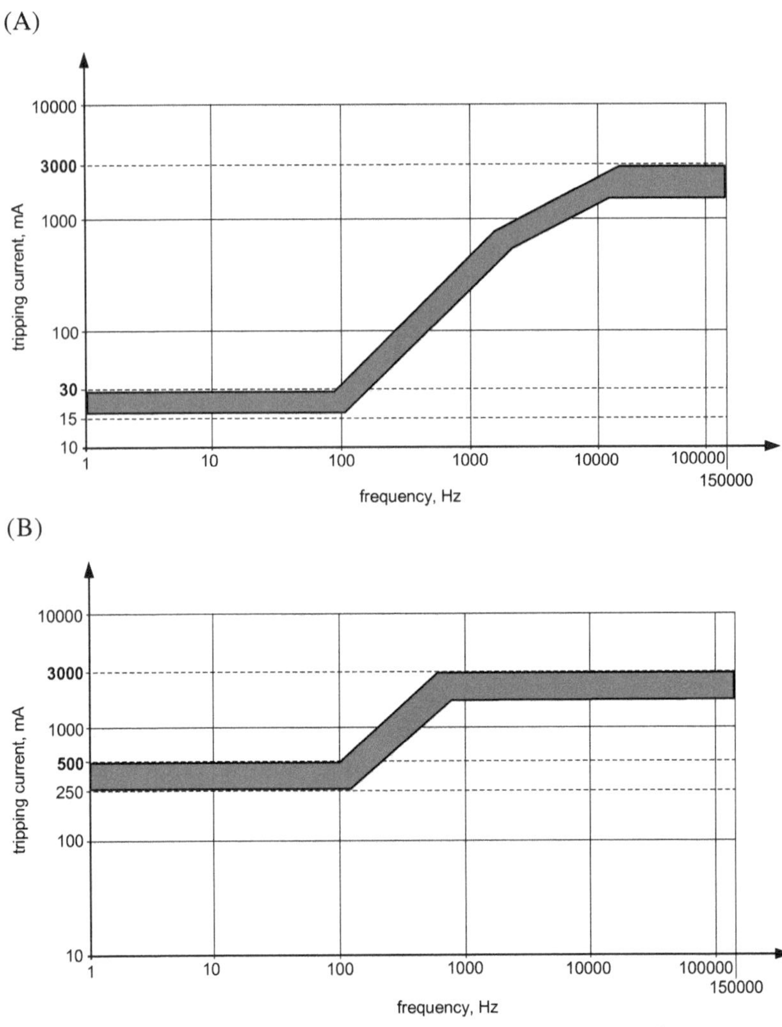

Fig. 8.26 Declared tripping characteristics of the RCDs within the frequency range 0–150 kHz: (A) 30 mA RCD according to [17], (B) 500 mA RCD according to [18].

Fig. 8.29 presents the results of the laboratory test of the tripping current of the selected two RCDs. According to the standard [12], for sinusoidal waveform the tripping threshold of residual current devices should be within the range of $(0.5–1.0)I_{\Delta n}$, where $0.5I_{\Delta n}$ is the rated residual nonoperating current. The same criterion range has been assumed in the test. The tripping threshold (rms value) should be neither equal to and lower than $0.5I_{\Delta n}$ nor higher than $1.0I_{\Delta n}$ (hatched area in Fig. 8.29). The tripping threshold of

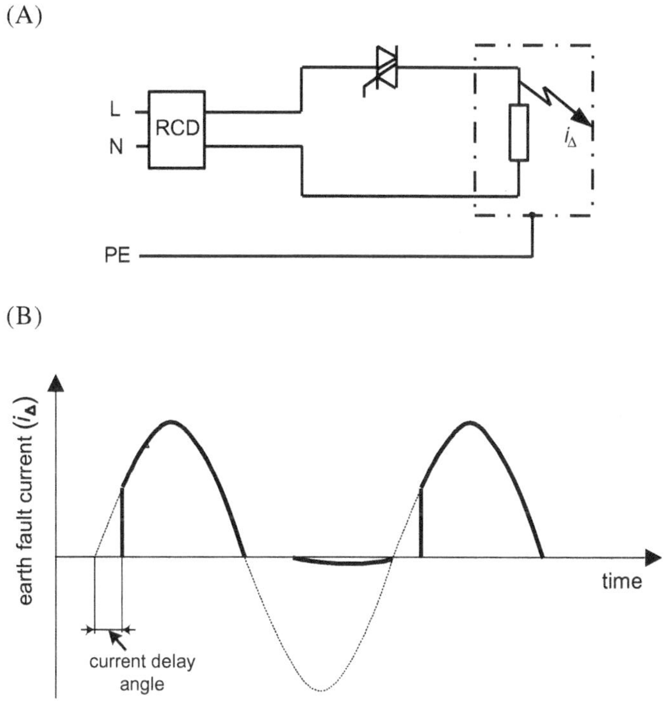

Fig. 8.27 Simplified circuit with converter for current delay angle control (A) and earth fault current waveform i_Δ in such a circuit (B).

RCDs lower than $0.5I_{\Delta n}$ can lead to their nuisance tripping. Results of the laboratory test show that the higher current delay angle, the lower is the tripping threshold of the tested RCDs. Moreover, when the current delay angle is 135°, the tripping current of A-type no-delayed tested RCD (Fig. 8.29B) is under the assumed permissible limit ($0.5I_{\Delta n}$). On the basis of this analysis, it can be concluded that earth fault current characterized by current delay angle, even strongly distorted (high value of the delay angle) is very easy to detect by RCDs. There is no risk of ineffective protection against electric shock, but there is a slight risk of nuisance tripping of RCDs.

8.5 Recommendations for residual current devices application to circuits with nonsinusoidal earth fault currents

Application of RCDs to circuits characterized by nonsinusoidal currents can be performed on the basis of Table 8.2. AC-type residual current devices are

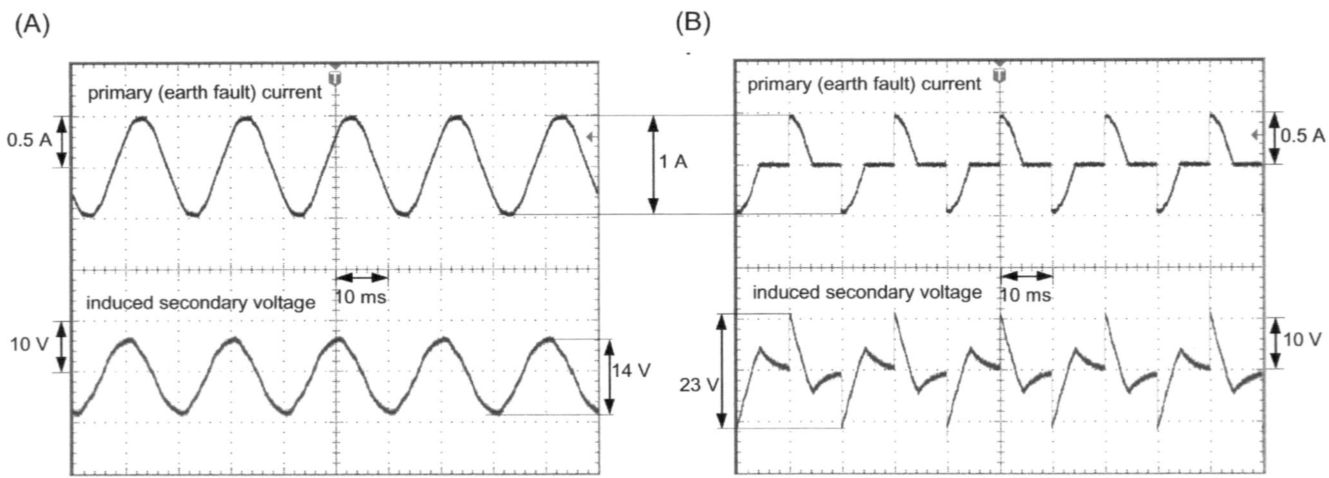

Fig. 8.28 Oscillograms of the induced secondary voltage in the current transformer of a 300 mA A-type RCD. Current delay angle of the primary (earth fault) current: (A) 0°, (B) 90°.

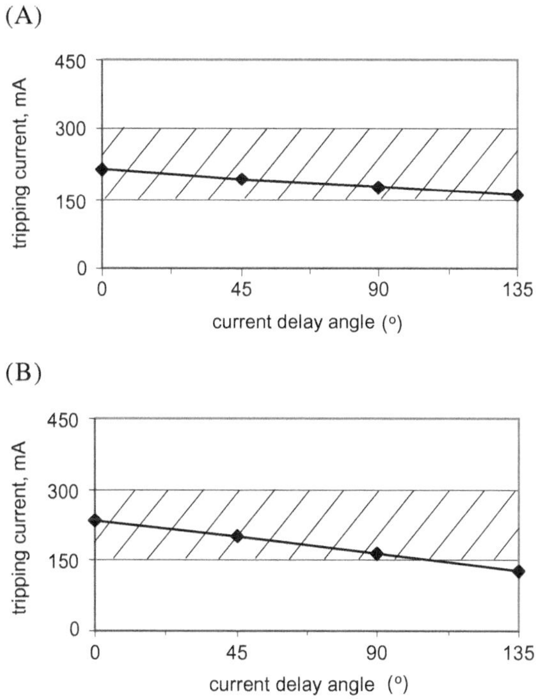

Fig. 8.29 Tripping current (rms value) of two selected 300 mA RCDs: (A) AC-type, delayed, (B) A-type, no-delayed.

sufficient for alternating residual currents. Note that they are also suitable in circuits with symmetrical phase angle control. The possibility of utilization of AC-type RCDs determines the fact that the residual current is bidirectional. Obviously, the frequency of the current should be of a power network frequency (50/60 Hz). A-type RCDs are allowed to be used in the same circuit as AC-type RCDs, and moreover, in rectifier circuits, but supplied L-N. The only exception is a single-phase rectifier circuit with smoothing (No. 8 in Table 8.2). In such a circuit, the residual current has too low pulsation to be detected by A-type RCDs. F-type RCDs have possibilities for detection of residual currents the same as A-type RCDs, and residual currents in circuits with single-phase supplied frequency converters. In frequency converter circuits earth fault currents with harmonics are suspected, so it excludes AC-type and A-type RCDs. The most technically advanced are B and B+ type RCDs. They are required in circuits, in which smooth direct current may flow, and/or earth fault current is composed of high-frequency components.

Table 8.2 Residual current waveforms i_Δ in converter circuits and suitable RCDs; according to own analysis and recommendations from [11,20,21].

No.	Converter type	Simplified diagram	Residual current waveform	Suitable RCD types
1	No converter	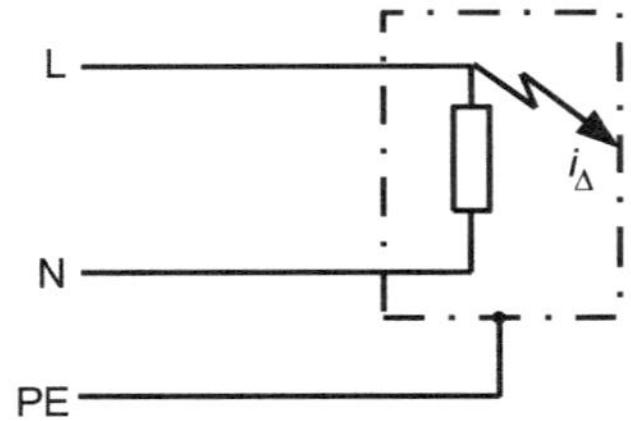	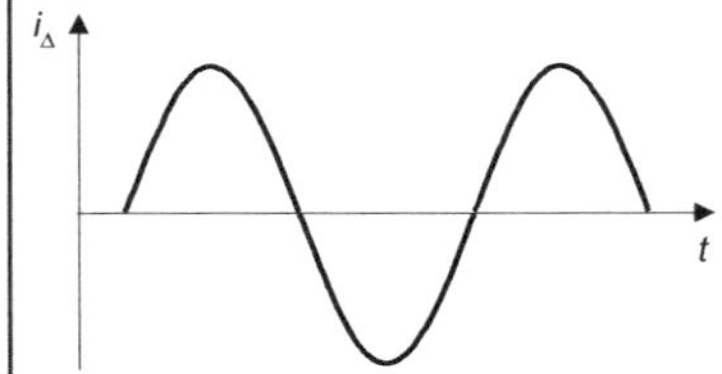	AC, A, F, B, B+
2	Symmetrical phase control	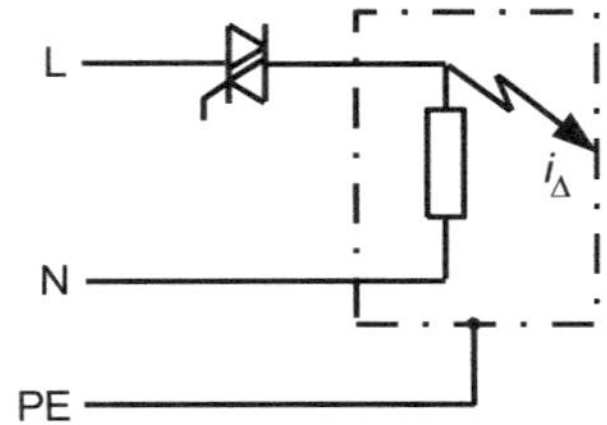	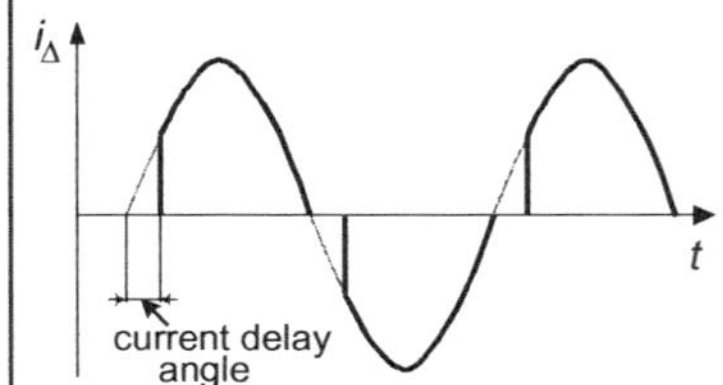	AC, A, F, B, B+
3	Burst control	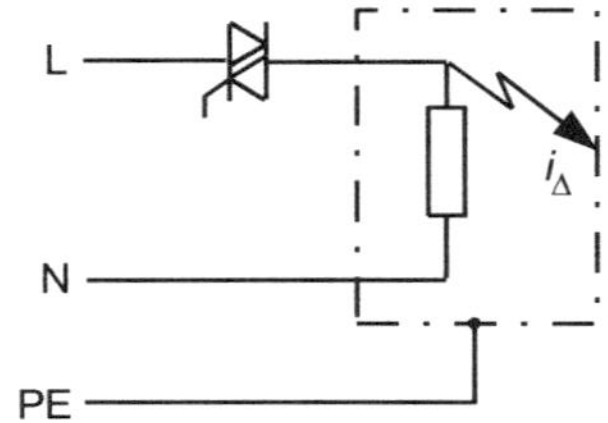	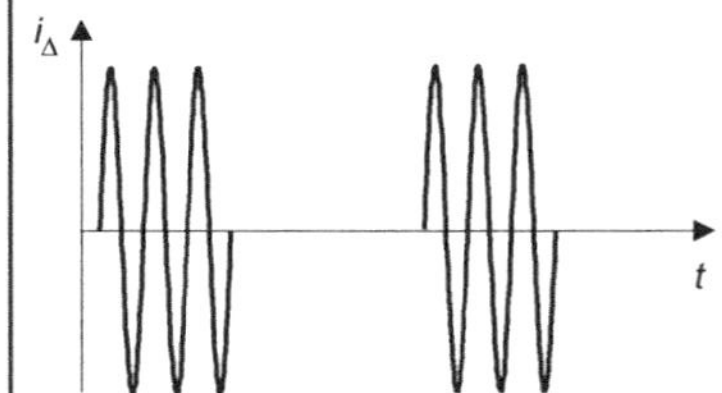	AC, A, F, B, B+

4	Half-wave rectifier	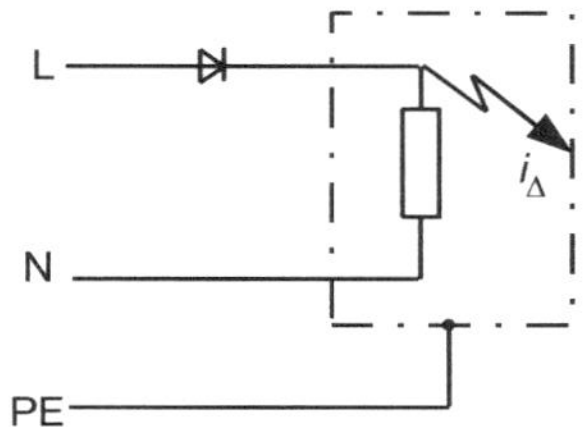	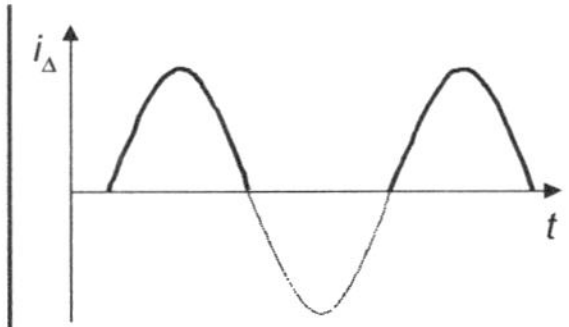	A, F, B, B+
5	Two-pulse bridge supplied L-N	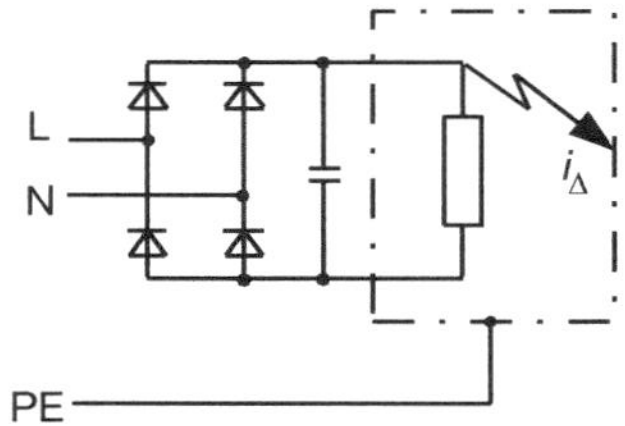	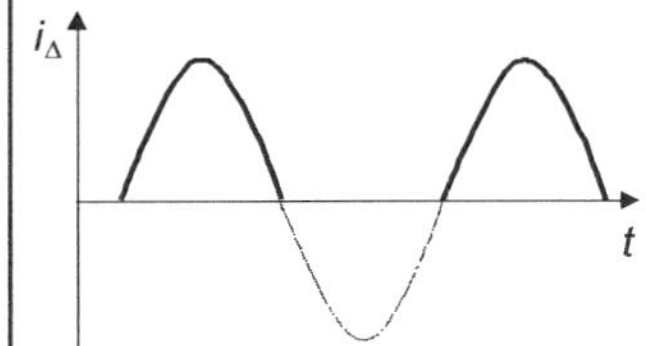	A, F, B, B+
6	Full-wave rectifier supplied L-N, half controlled	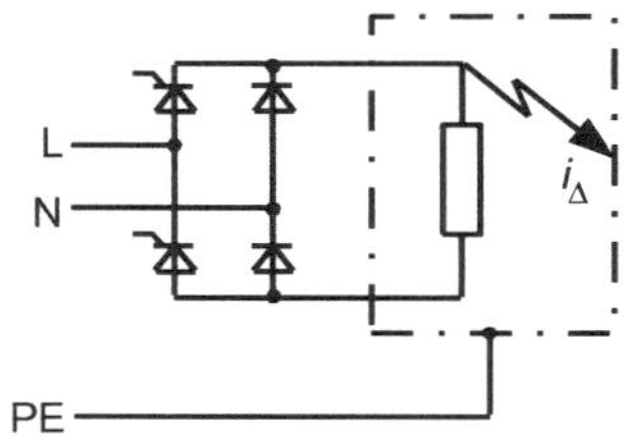	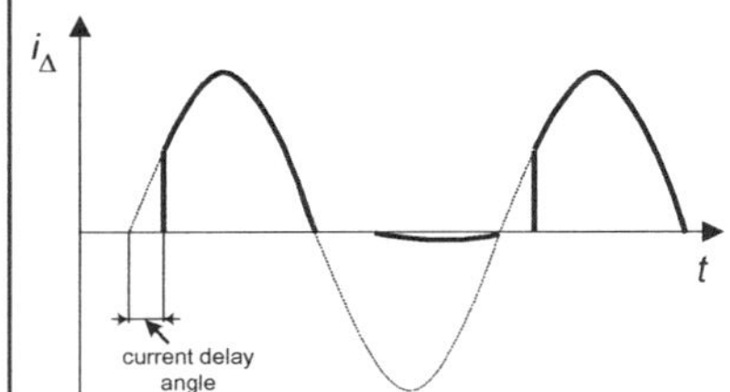	A, F, B, B+
7	Frequency converter (FC) supplied L-N	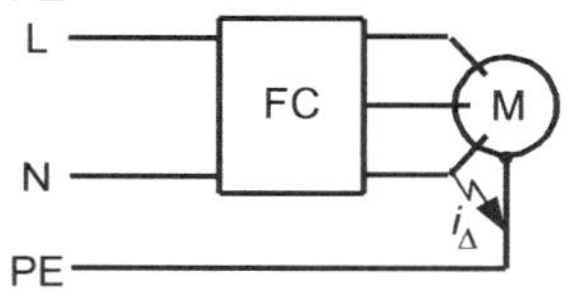	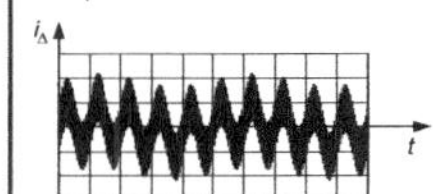	F, B, B+

Continued

Table 8.2 Residual current waveforms i_Δ in converter circuits and suitable RCDs; according to own analysis and recommendations from [11,20,21]—cont'd

No.	Converter type	Simplified diagram	Residual current waveform	Suitable RCD types
8	Half-wave rectifier with smoothing	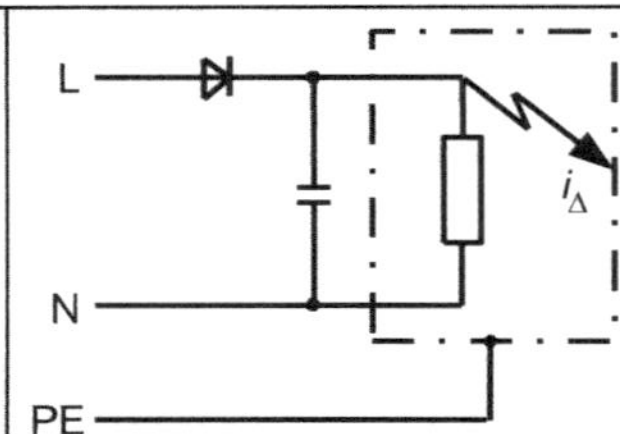	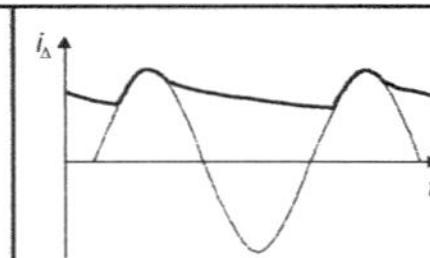	B, B+
9	Two-pulse bridge supplied from two phases	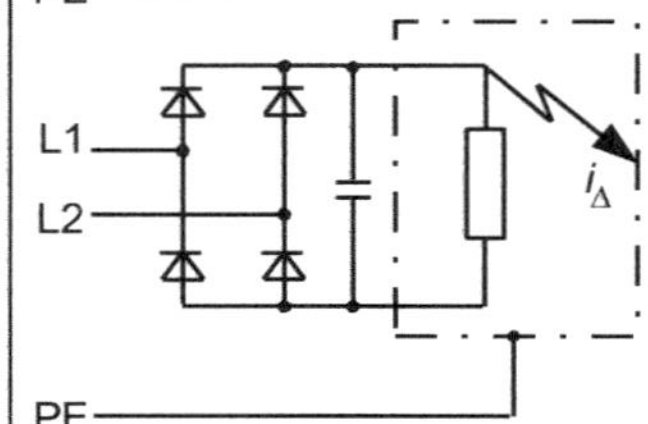	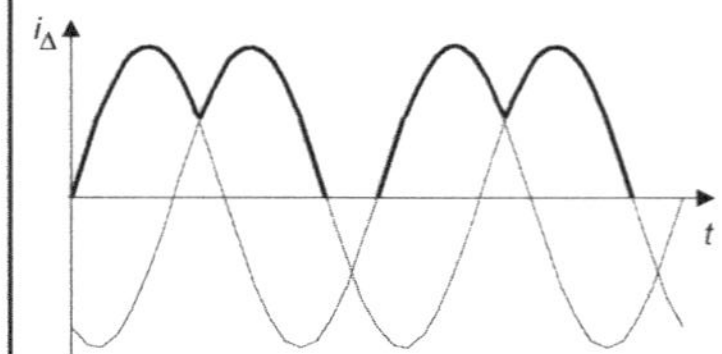	B, B+
10	Three-phase rectifier bridge	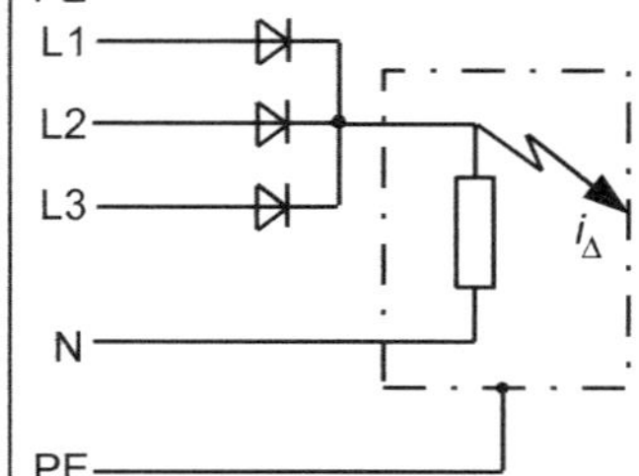	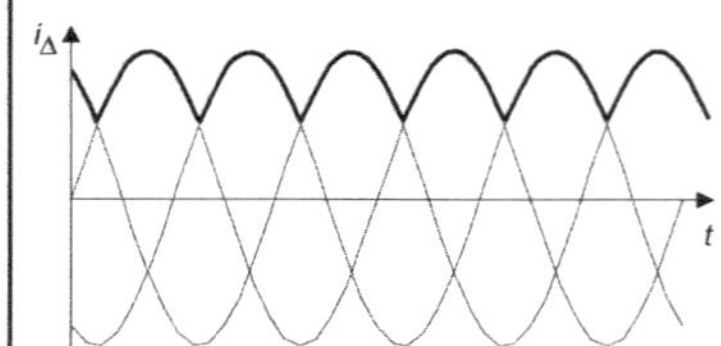	B, B+

11	Six-pulse bridge	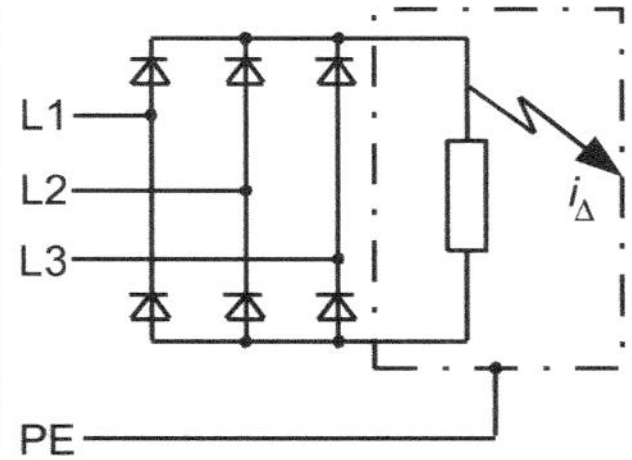	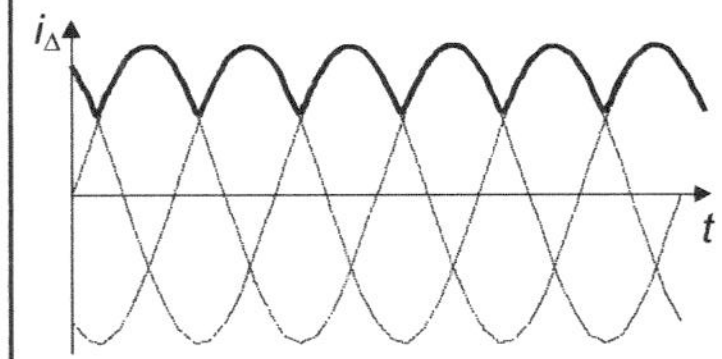	B, B+
12	Frequency converter (FC) supplied from three phases	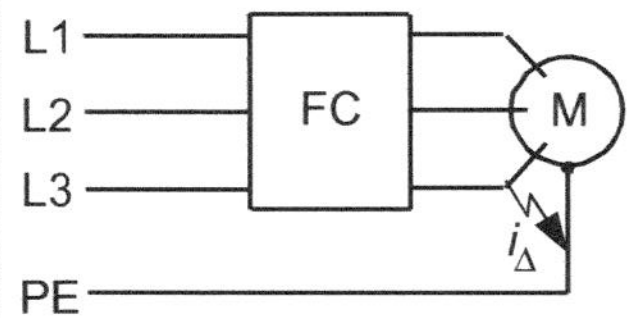	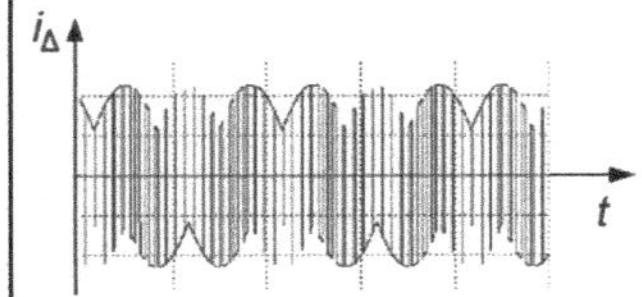	B, B+

References

[1] A.M. Featherstone, A.S. Sastrosubroto, An examination of RCD performance during system disturbances (and other relays), in: IEE Colloquium on Safeguarding Industrial Plant During Power System Disturbances, London, UK, 4-4 Dec, 1989.
[2] T.M. Lee, T.W. Chan, The effects of harmonics on the operational characteristics of residual current circuit breakers, in: International Conference on Energy Management and Power Delivery, Proceedings of EMPD'9, vol. 2, Nov. 1995, pp. 715–719.
[3] D.Y. Luo Xiang, X.H. Wang, M.L. Chen, Tripping characteristics of residual current devices under nonsinusoidal currents, IEEE Trans. Ind. Appl. 3 (47) (2011) 1515–1521.
[4] Y. Wang, K. Li, C. Liu, Z. Guo, F. Niu, X. Zhang, Study on modeling and simulation of AC/DC sensitive residual current transformer, in: 2011 1st International Conference on Electric Power Equipment—Switching Technology, Xi'an, China, 23–27 Oct., 2011.
[5] F. Freschi, High-frequency behavior of residual current devices, IEEE Trans. Power Delivery 3 (27) (2012) 1629–1635.
[6] Y. Han, C. Ding, X. Shou, Design and implementation of an A-type residual current circuit breaker IC, in: 2012 IEEE International Symposium on Industrial Electronics, Hangzhou, China, 28–31 May 2012, 2012.
[7] T. Wieland, M. Aigner, E. Schmautzer, J. Pasker, L. Fickert, Influences on safety issues for inverter supplied grid structures, in: 2012 Electric Power Quality and Supply Reliability, Tartu, Estonia, 11–13 June 2012, 2012.
[8] P. Czaja, Examination of the impact of design of a residual current protective device on the release frequency range, in: 2017 Progress in Applied Electrical Engineering (PAEE), Koscielisko, Poland, 25–30 June 2017, 2017.
[9] S. Czapp, K. Dobrzyński, J. Klucznik, Z. Lubośny, R. Kowalak, Improving sensitivity of residual current transformers to high frequency earth fault currents, Arch. Elect. Eng. 66 (3) (2017) 485–494.
[10] S. Czapp, The effect of PWM frequency on the effectiveness of protection against electric shock using residual current devices, in: 2010 International School on Nonsinusoidal Currents and Compensation (ISNCC), Łagów, Poland, 15–18 June 2010, 2010.
[11] HD 60364-5-53:2015, Low-voltage electrical installations—Part 5-53: Selection and erection of electrical equipment—Switchgear and controlgear, 2015.
[12] IEC 61008-1:2010, Residual current operated circuit-breakers without integral overcurrent protection for household and similar uses (RCCBs)—Part 1: General rules, 2010.
[13] IEC 61557-6:2007, Electrical safety in low voltage distribution systems up to 1000 V a.c. and 1500 V d.c.—Equipment for testing, measuring or monitoring of protective measures—Part 6: Effectiveness of residual current devices (RCD) in TT, TN and IT systems, 2007.
[14] EN 50178:1997, Electronic equipment for use in power installations, 1997.
[15] J. Schoneck, Y. Nebon, LV protection devices and variable speed drives, Cahier technique no. 204. Schneider Electric, 2002.
[16] G. Grünebast, Allstromsensitive Fehlerstromschutzeinrichtungen. Teil 2: Vorschriftsmässiger Einsatz, Elektropraktiker 2 (62) (2008) 144–149.
[17] DFS 2 016-2/0,03 B SK, Data Sheet, Doepke, 2019.
[18] DFS 4 016-4/0,5 B SK HD, Data Sheet, Doepke, 2019.
[19] S. Czapp, J. Horiszny, The effect of current delay angle on tripping of residual current devices, in: 2017 International Conference on Information and Digital Technologies (IDT), Zilina, Slovakia, 5–7.07.2017, 2017, pp. 90–94.
[20] Residual Current Devices, Application guide, Eaton, 2017.
[21] Doepke, Residual Current Protection & Selection for UK Installations, Doepke UK Ltd, 2017.

CHAPTER 9

Earth fault loop impedance measurement in circuits with residual current devices

9.1 General information

Every electrical installation is verified initially, before being put into service, as well as periodically, during its utilization. The main reason for verification is to check whether protection against electric shock is on the satisfying level. According to standard IEC 60364–6 [1], the following tests are recommended to be performed:

- continuity of conductors,
- insulation resistance measurement,
- insulation resistance testing to confirm the effectiveness of protection by SELV, PELV, or electrical separation,
- insulation resistance testing to confirm the effectiveness of floor and wall resistance/impedance,
- polarity,
- effectiveness of automatic disconnection of supply,
- effectiveness of additional protection,
- phase sequence,
- functional tests,
- voltage drop.

In case of verification of the effectiveness of automatic disconnection of supply, earth fault loop impedance calculations or measurements are carried out [1–4], especially in TN low-voltage systems. In such systems, the measured value of the earth fault loop impedance Z_s is the base for evaluation of the effectiveness of protection against electric shock by automatic disconnection of supply, according to formula (4.1).

Fig. 9.1 presents a circuit during the earth fault loop impedance measurement in a system with an RCD. In practice, this measurement is performed

Residual Current Devices
https://doi.org/10.1016/B978-0-323-89783-9.00002-3

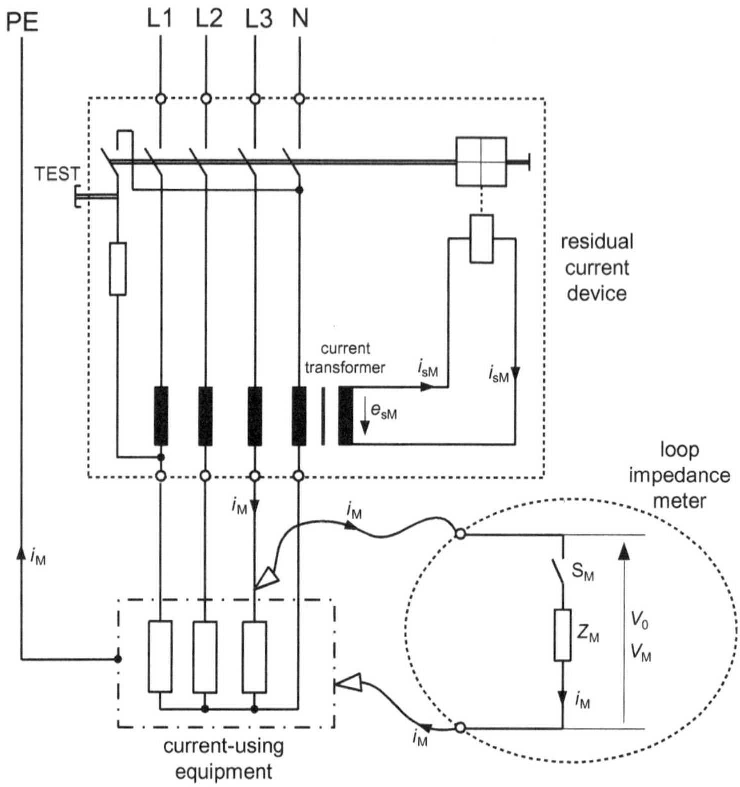

Fig. 9.1 Principle of the earth fault loop impedance measurement in circuit with a residual current device. S_M—switch, Z_M—impedance of the meter, i_M—measuring current, i_{sM}—current in the secondary circuit of the current transformer, e_{sM}—induced voltage in the secondary winding of the current transformer, V_0—measured voltage before closing switch S_M, V_M—measured voltage after closing switch S_M.

with the use of the so-called artificial short-circuit method [5–17]. After connection of the loop impedance meter to the exposed-conductive-part (metallic enclosure) of the current-using equipment, V_0 voltage is measured (switch S_M is opened). This voltage is recorded in the meter memory and then V_M voltage is measured (switch S_M is closed). Earth fault loop impedance Z_s is evaluated according to the following expression:

$$Z_s = \frac{\Delta V}{I_M} = \frac{V_0 - V_M}{\frac{V_M}{Z_M}} = Z_M\left(\frac{V_0}{V_M} - 1\right) \approx |Z_M|\left(\frac{|V_0|}{|V_M|} - 1\right) \quad (9.1)$$

where

Z_s is the measured earth fault loop impedance,

Z_0 is the meter internal impedance,

V_0 is the voltage measured before closing switch S_M,

V_M is the voltage measured after closing switch S_M,

I_M is the measuring current.

One can observe that the measuring current i_M (I_M) flows through the PE conductor, and it is a residual current for the RCD installed upstream the point of the measurement. The RCD interprets this current as an earth fault current or body current, and unwanted tripping of this RCD may occur. If so, the earth fault loop impedance measurement cannot be finished properly.

It turns out that the measurement of the earth fault loop impedance in circuits with RCDs, without their unwanted tripping, is possible in some cases, provided that the value and time of the measuring current flow are recognized. It is also important which type of RCD is installed: AC-type, A-type, short-time delayed, or delayed (S-type, selective). From the point of view of the earth fault loop impedance measurement, the main factor is a delay in the tripping of the RCD in the circuit. As described in Chapter 3, the following are the types of RCDs [18–20]:

- no-delayed,
- short-time delayed,
- delayed (S-type, selective),
- differential relay—composed of an external current transformer and a circuit-breaker, it is possible to set both the delay time and the tripping current.

Each type of the aforementioned RCDs has the tripping characteristic, including delay, specified by standards or manufacturers. The knowledge about the tripping characteristic enables to evaluate the behavior of the RCD during the measuring current flow.

9.2 Properties of the loop impedance meters and testing of residual current devices

In order to evaluate whether a particular RCD opens the tested circuit during the earth fault loop impedance measurement, it is necessary to recognize the time of measuring current flow in the meter used for the test. The most popular meters use either half-wave current (10 ms in 50 Hz power network) or full-wave current ($\geq$20 ms in 50 Hz power network). The first type of meter (of half-wave current) is usually able to measure only resistance, not impedance.

Fig. 9.2 presents an oscillogram of the measuring current i_M for a selected meter commonly used by practitioners. This meter (Meter 1) uses half-wave current (10 ms) of peak value approx. 30 A and measures only earth fault loop resistance. It is acceptable in final circuits of electrical installations, because, due to the small cross-section of conductors, their resistance dominates there.

A full-wave current uses the meter (Meter 2) for which the oscillogram of the measuring current waveform i_M is presented in Fig. 9.3. This current flows within 30 ms and its peak value is also approx. 30 A. Meter 2 enables for measuring the earth fault loop impedance, as well as its orthogonal components: resistance and reactance.

An analysis of the typical measuring current waveforms (Figs. 9.2 and 9.3) enables to conclude that in the case of meters that utilize half-wave current it is possible to measure earth fault loop impedance (resistance) when in the tested circuit a short-time delayed RCD (10 ms delay), a delayed (S-type) RCD, or a delayed differential relay is installed. They do not react to half-wave current, regardless of whether their type is AC or A. Meter 2, forcing measuring current within 30 ms, should not cause unwanted tripping of delayed (S-type) RCDs or delayed differential relays. S-type RCDs have a delay time equal to 40 ms, which is more than the measuring current flow. Additionally, it will not cause a reaction of delayed differential relays, whose delay time is usually higher than 40 ms, in order to keep selective operation with S-type RCDs. Thus the problem of unwanted tripping is especially expected in the case of no-delayed RCDs.

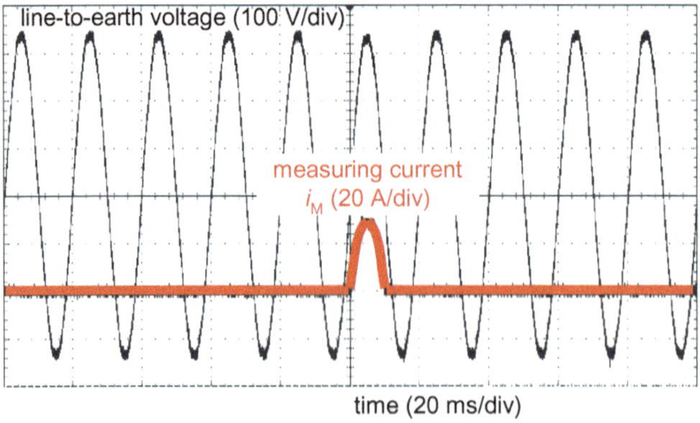

Fig. 9.2 Measuring current waveform of the Meter 1.

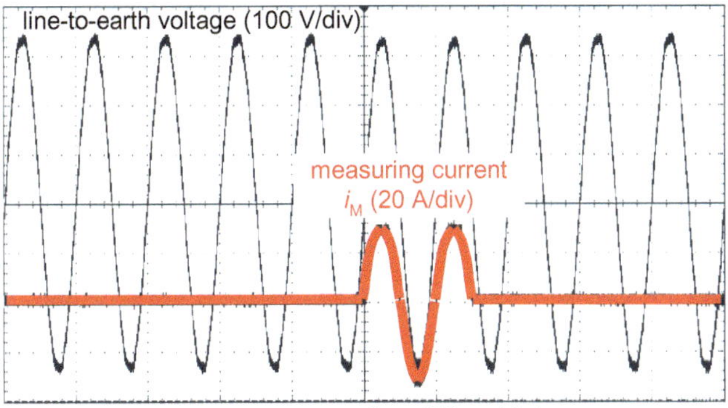

Fig. 9.3 Measuring current waveform of the Meter 2.

To verify the aforementioned observation, a laboratory test of RCDs has been performed, with the use of the meters forcing currents presented in Figs. 9.2 and 9.3. During this test, various types of RCDs (AC-type, A-type, short-time delayed, S-type, 30 mA and 300 mA) were installed in the circuit, and then the earth fault loop impedance measurement was performed. Each selected RCD was tested 3 times—time after time with intervals of about 1–2 s. In the case of the half-wave measuring current (Meter 1), separate tests were performed for the positive polarity of the half-wave current and the negative polarity of the half-wave current.

Fig. 9.4 presents representative results of the test. In this figure, three results (trial 1, trial 2, and trial 3 performed with intervals of about 1–2 s) of the "tripping"/"no tripping" of the tested RCD for each measuring current waveform are included.

The first residual current device, marked RCD-A1-30mA (rated residual operating current 30 mA, A-type, no-delayed), tripped out during every earth fault loop impedance measurement. It can be observed in Fig. 9.4A as red bars with the annotation "tripping." It is a very unfavorable case from the point of view of accurate measurement of the earth fault loop impedance. Performance of the proper measurement is impossible, with the use of the typical meters.

Slightly different behavior was recorded in the case of the test of RCD-AC1-30mA (rated residual operating current 30 mA, AC-type, no-delayed)—Fig. 9.4B. For half-wave current, tripping of the RCD was observed only during the first trial (trial 1). For the other two trials, no tripping occurs. This phenomenon is related to the magnetic characteristic of the iron core of the current transformer included in the tested AC-type

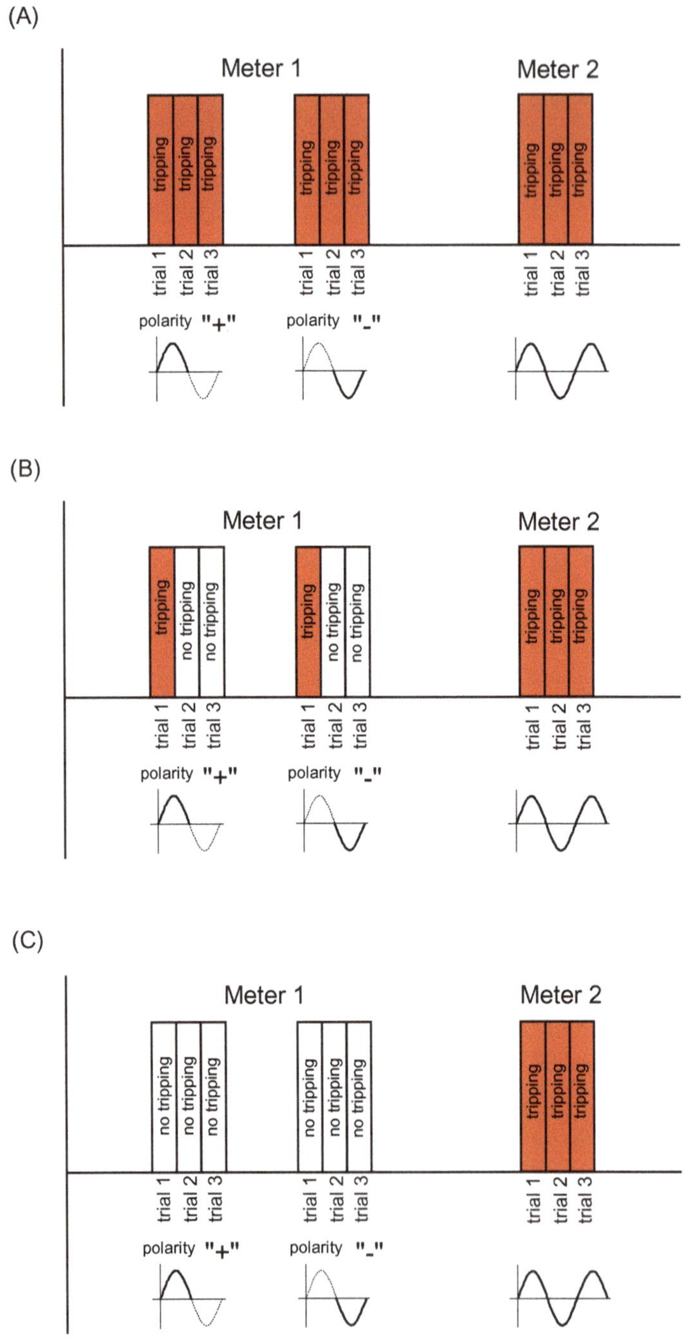

Fig. 9.4 The behavior of the RCDs during earth fault loop impedance measurement: (A) RCD-A1-30mA: rated residual operating current 30 mA, A-type, no-delayed, (B) RCD-AC1-30mA: rated residual operating current 30 mA, AC-type, no-delayed, (C) RCD-AC2-30mA: rated residual operating current 30 mA, AC-type, short-time delayed (10 ms),

(Continued)

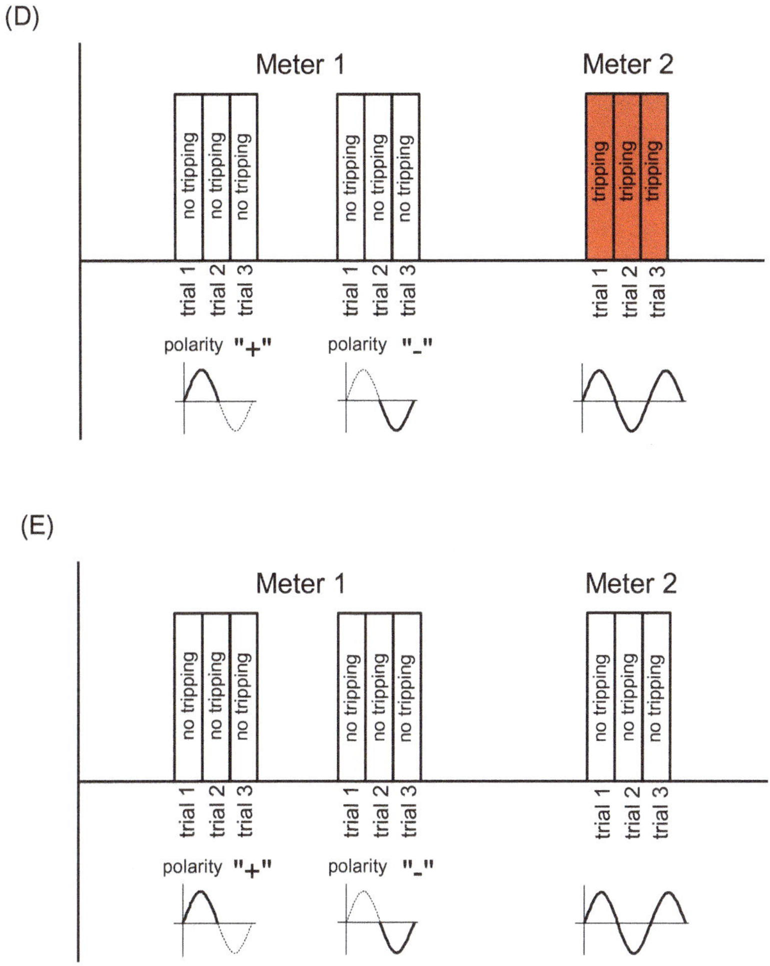

Fig. 9.4—cont'd (D) RCD-A2-30mA: rated residual operating current 30 mA, A-type, short-time delayed (10 ms), (E) RCD-A3-300mA: rated residual operating current 300 mA, A-type, delayed (S-type, 40 ms).

RCD. As presented in Fig. 2.5, there are three conventional hysteresis loops of the RCD current transformer cores. An AC-type RCD is the simplest one and it is devoted to the detection of 50/60 Hz sine residual currents only. For such a purpose, the rectangular hysteresis loop is sufficient. However, this type of iron core is not suitable for the detection of the half-wave current—as presented in Fig. 9.5A.

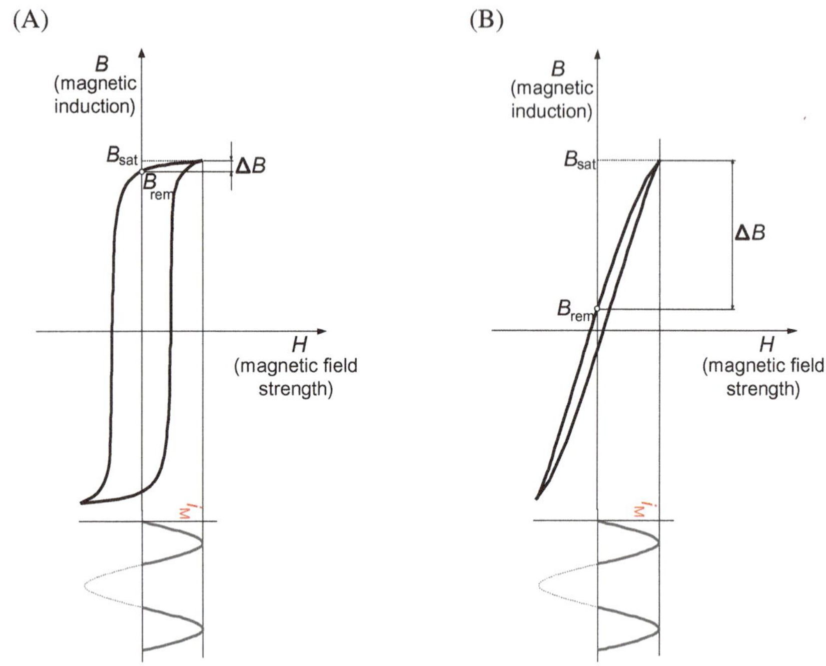

Fig. 9.5 Two types of the hysteresis loop: (A) rectangular (high ratio B_{rem}/B_{sat}), (B) flat (low ratio B_{rem}/B_{sat}). B_{sat}—saturation magnetic induction, B_{rem}—remanence magnetic induction, ΔB—a variation of magnetic induction in case of half-wave measuring current i_M.

The half-wave current gives a too low variation of magnetic induction ΔB, to induce a sufficient value of the voltage in the secondary winding of the current transformers (see induced voltage $e_s(t)$ in Fig. 2.6B). Only an iron core of a flat hysteresis loop should be used (Fig. 9.5B)—a variation of magnetic induction ΔB is relatively high then. This justifies the behavior (no tripping) of RCD-AC1-30mA in the case of trial 2 and trial 3. The iron core of the current transformer is saturated after the first trial (trial 1). During trial 1 (just after changing the polarity of the measuring current) tripping of this RCD occurs, because the iron core is demagnetized and variation of the magnetic induction for this first trial (the first half-wave) is sufficient to trip out the RCD. Forcing a sinusoidal measuring current (Meter 2) gives nuisance tripping of the RCD-AC1-30mA in every trial.

Application of short-time delayed RCDs

- RCD-AC2-30mA: rated residual operating current 30 mA, AC-type, short-time delayed (10 ms), Fig. 9.4C,
- RCD-A2-30mA: rated residual operating current 30 mA, A-type, short-time delayed (10 ms), Fig. 9.4D

enables for the earth fault loop impedance (resistance) measurement with the use of Meter 1 (half-wave). It is convergent with the previous theoretical consideration—measuring current flows within the time 10 ms, what is the guaranteed delay in tripping by short-time delayed RCDs. Unfortunately, using full-wave current (Meter 2) results in the unnecessary operation of both mentioned tested RCDs.

Only S-type RCDs (Fig. 9.4E) are resistant to the measuring current used in the laboratory test. It was expected because its guaranteed delay in tripping is higher than the time flow of every forced measuring current.

Taking into account the previously mentioned laboratory test and considerations, the conclusion is that in the case of circuits with some types of RCDs (short-time delayed, delayed), the measurement of earth fault loop impedance is possible even with the use of conventional meters characterized by relatively high measuring current 10–30 A. Difficulties in measurements may occur in circuits equipped with no-delayed A-type RCDs. For such circuits, special methods should be utilized.

9.3 Special methods of the measurement to circuits with residual current devices

9.3.1 Very low-current method

Earth fault loop impedance measurement in circuits protected by RCDs can be easily performed with the use of the measuring current lower than the rated residual nonoperating current of the RCD. The most commonly used RCDs have a rated residual operating current equal to 30 mA because such devices are required in certain circuits/locations, which is explained in Chapter 6. Unwanted tripping of 30 mA RCDs does not occur if a measuring current has a value not higher than 15 mA ($0.5I_{\Delta n}$). Many meters available on the market have just that value of the measuring current (Fig. 9.6).

However, it should be underlined that the low value of the measuring current may not give sufficient accuracy of the measurement. Fig. 9.7 presents the earth fault loop impedance measurement error, as a function of the argument of the tested circuit impedance, for three values of measuring current. One can see that the higher value of the measuring current, the higher is the measurement accuracy (lower measurement error).

In some manuals of the earth fault loop impedance meters, it is written that an option with a very low value of the measuring current (usually 10–15 mA) may be selected only for measurement of the loop impedance not lower than 0.4–0.5 Ω. If the measured impedance has a lower value, the total error of the measurement may exceed 30%, which, according to

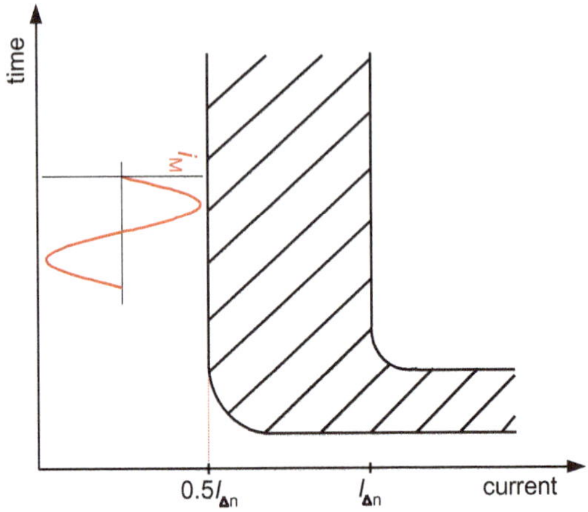

Fig. 9.6 Principle of the use of measuring current i_M lower than the rated residual nonoperating current of RCDs ($0.5I_{\Delta n}$).

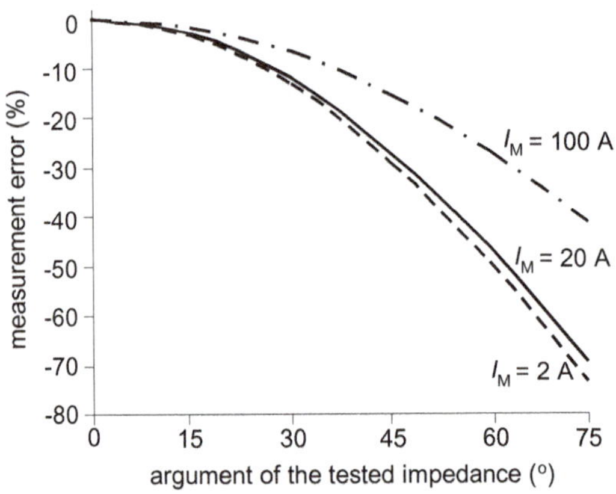

Fig. 9.7 Error of the earth fault loop impedance measurement as a function of the argument of the tested impedance, for various values of the measuring current I_M (argument of the meter: $\phi_M = 0°$—resistance).

standard EN 61557 [21], is specified as the maximum permissible percentage operating uncertainty. Alternatively, other special methods of measurements, dedicated to circuits with RCDs, should be used [22–27].

9.3.2 Method with a compensating current

This method utilizes additional current flow, whose aim is to compensate the measuring current. In consequence, the magnetic flux in the current

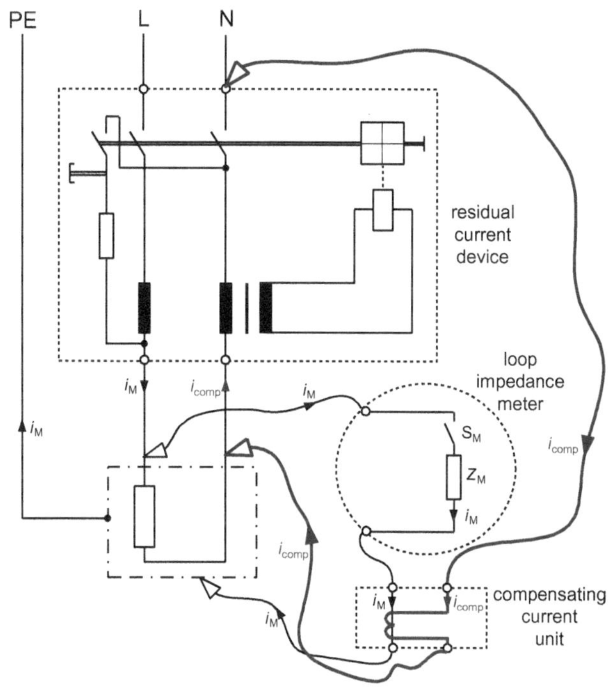

Fig. 9.8 Principle of the earth fault loop impedance measurement with the method using a compensating current i_{comp}; for other symbols see Fig. 9.1

transformer of the residual current device is balanced. Fig. 9.8 presents the principle of the measurement according to this method [25].

In this method, the earth fault loop impedance meter is connected (or integrated) with an additional unit—a compensating current unit. The impedance meter forces measuring current i_M in a classical way, after closing switch S_M. This current flows from line conductor L across the line terminal of the RCD, the loop impedance meter, and returns to the supplying network via the PE conductor. Simultaneously, the compensating current i_{comp} flows across the neutral (N) path of the RCD in the opposite direction to the measuring current i_M, which flows across the line (L) path of the RCD. If the geometric sum of these currents $i_M + i_{comp}$ is close to 0, there is no risk of unwanted tripping of the RCD. In practice, the permissible value of this geometric sum depends on the rated residual operating current of the RCD in the tested circuit. For RCDs of $I_{\Delta n} = 30\,\text{mA}$ the sum should not exceed 15 mA ($0.5 I_{\Delta n}$).

The main element in the compensating current unit is a transducer of the rated current ratio equal to 1:1. This unit can be integrated with the earth fault loop impedance meter or can be a separate device. An unfavorable feature of this method is the necessity of connection of the one terminal of the compensating current unit to the neutral conductor upstream of the RCD. When the point of the measurement (e.g., socket-outlet) is far from a distribution board with RCDs, the additional conductor connected to this neutral conductor is long.

9.3.3 Saturation of the residual current device current transformer

Tripping of a residual current device occurs when the residual current has sufficient value and is able to produce a sufficient level of the induced voltage e_{sM} in the secondary winding of the current transformer of RCDs.

However, if the iron core of the current transformer is saturated, then the induced voltage e_{sM} can be negligible, in spite of the high value of the residual (measuring) current. Fig. 9.9 presents the idea of the measurement with the saturation of the current transformer [26].

Before closing switch S_M (before forcing measuring current i_M), the DC unit produces slowly rising direct (saturating) current i_{DC}. The rising of the saturating current value has to be performed slowly in order not to generate a high variation of magnetic induction in the current transformer. After stabilizing the value of this current (current transformer is saturated), the measuring current i_M is superimposed on the saturating DC component (i_{DC})—Fig. 9.10.

The DC component has to be high enough to ensure saturation of the current transformer during the measuring current flow. It is recommended that the DC component in a steady state is higher than peak-to-peak value of the measuring current.

After finishing the measurement (the measuring current is not flowing), the DC component has to be decreased slowly (similarly, as during its rising). This method is patented [26] and its details are presented in the paper [22].

The effect of saturation of the current transformer of RCDs is utilized in the meter available on the market [27]. However, the main difference with reference to the previously presented method is the use of half-wave measuring current only. Fig. 9.11 presents an oscillogram of the measuring current for this meter. The measuring half-wave current (28 A in peak) is preceded by half-wave saturating current impulses which are forced permanently in every approx. 20 ms, in order to keep the current transformer of the RCD in the saturation state. All half-wave impulses (saturating and measuring) have to have the same polarization. Otherwise, unwanted tripping of the RCD may occur.

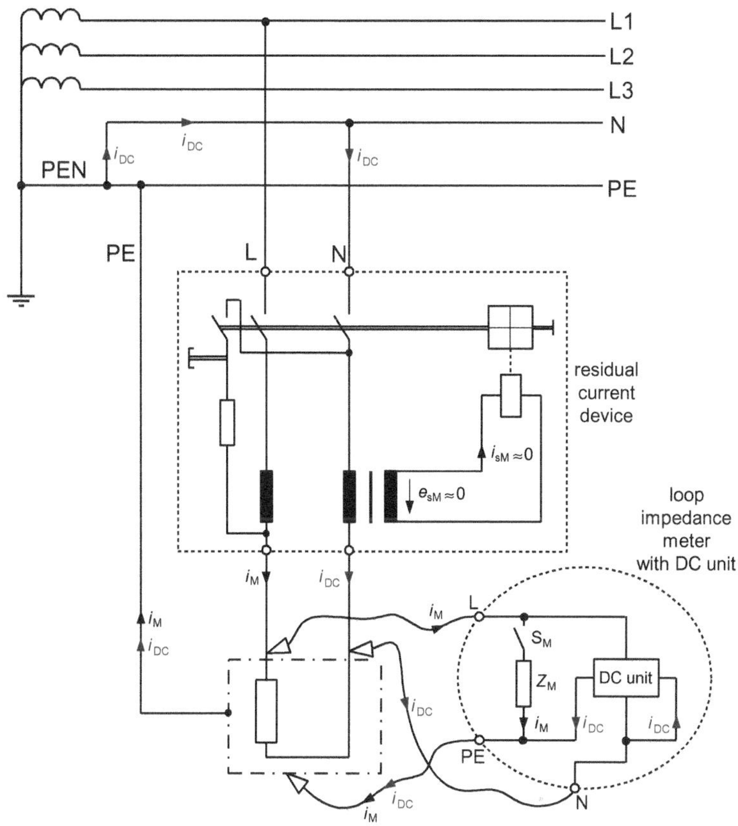

Fig. 9.9 Earth fault loop impedance measurement with the saturation of the RCD current transformer, i_{DC}—direct (saturating) current; for other symbols see Fig. 9.1.

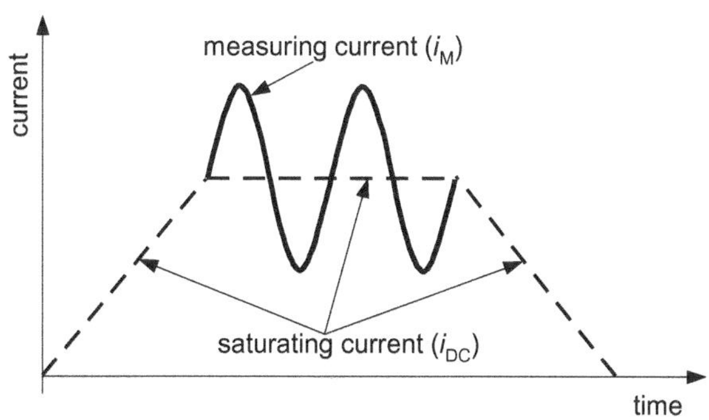

Fig. 9.10 Coordination of the saturating current i_{DC} with the measuring current i_M.

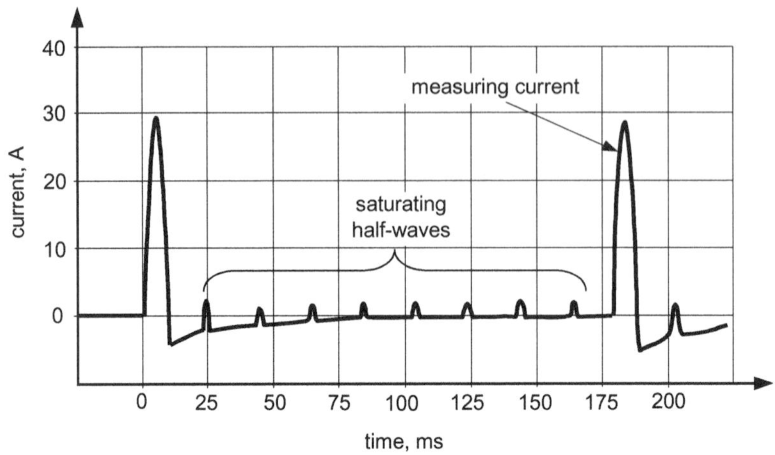

Fig. 9.11 Waveform of the measuring current for meter KEW4120A; according to [27].

References

[1] IEC 60364-6:2016, Low-voltage electrical installations—Part 6: Verification, 2016.
[2] H. Liu, M. Mitolo, J. Qiu, Ground-fault loop impedance calculations in low-voltage single-phase systems, IEEE Trans. Ind. Appl. 2 (50) (2014) 1331–1337.
[3] A. Wells, Initial verification, inspection and testing—an electrical contractor's view, in: IET Seminar on the Impact of the 17th Edition of the IEE Wiring Regulations, 2008.
[4] HD 60364-4-41:2017, Low-voltage electrical installations—Part 4–41: Protection for safety—Protection against electric shock, 2017.
[5] R. Roskosz, A new method for measurement of earth fault loop impedance, IEEE Trans. Power Delivery 2 (6) (1991) 546–549.
[6] A.P.S. Meliopoulos, G. Cokkinides, H. Abdallah, S. Duong, S. Patel, A PC based ground impedance measurement instrument, IEEE Trans. Power Delivery 3 (8) (1993) 1095.
[7] M.B. Harris, A.W. Kelley, J.P. Rhode, M.E. Baran, Instrumentation for measurement of line impedance, in: Proc. of APEC'94, 1994, pp. 887–893.
[8] J.P. Rhode, A.W. Kelley, M.E. Baran, Complete characterization of utilization-voltage power system impedance using wideband measurement, IEEE Trans. Ind. Appl. 6 (33) (1997) 1472–1479.
[9] S. Czapp, R. Roskosz, A. Skiba, Vectorial method for measurement of short-circuit loop impedance in network with operating load, in: Proc. 16th IMEKO World Congress, Vienna, Sept., 2000.
[10] K.O.H. Pedersen, A.H. Nielsen, N.K. Poulsen, Short-circuit impedance measurement, IEE Proc. Gener. Transm. Distrib. 2 (150) (2003) 169–174.
[11] R. Roskosz, M. Ziolko, Measurement accuracy of short-circuit loop impedance in power systems, in: Proc. XVII IMEKO World Congr., TC4, Dubrovnik, June, 2003, pp. 903–907.
[12] E.E. Nino, A.B. Nassif, W. Xu, Network impedance measurement methods based on on-site data, in: Proc. of the 37th Annual North American Power Symposium, 2005, pp. 569–575.
[13] S. Czapp, Measurement of the earth fault loop impedance in a low voltage network with operating loads, Int. J. Power Energy Syst. 4 (29) (2009) 266–271.

[14] S. Czapp, Fault loop impedance measurement in low voltage network with residual current devices, Elektron. Elektrotech. 6 (122) (2012) 109–112.
[15] M. Kaspirek, D. Mezera, Voltage quality parameters in LV distribution grids in dependence, on short circuit impedance, in: 22nd International Conference on Electricity Distribution, Stockholm, June 2013, pp. 10–13.
[16] M. Aigner, E. Schmautzer, C. Sigl, Measurement of fault loop impedances in different low voltage network configurations, Electr. Eng. Electron. J. 1 (2013).
[17] M. Aigner, E. Schmautzer, C. Sigl, T. Wieland, L. Fickert, Fehlerschleifenimpedanz-Messung in Niederspannungsnetzen mit Wechselrichtern, in: 8. Internationale Energiewirtschaftstagung an der TU Wien, IEWT Wien, 13–15 Feb. 2013, 2013.
[18] IEC 60755:2017, General safety requirements for residual current operated protective devices, 2017.
[19] IEC 61008-1:2010, Residual current operated circuit-breakers without integral overcurrent protection for household and similar uses (RCCBs)—Part 1: General rules, 2010.
[20] IEC 61009-1:2010, Residual current operated circuit-breakers with integral overcurrent protection for household and similar uses (RCBOs)—Part 1: General rules, 2010.
[21] IEC 61557-3:2007, Electrical safety in low voltage distribution systems up to 1 000 V a. c. and 1 500 V d.c.—Equipment for testing, measuring or monitoring of protective measures—Part 3: Loop impedance, 2007.
[22] R. Roskosz, E. Musial, S. Czapp, A method of earth fault loop impedance measurement without unwanted tripping of RCDs, in: 2018 Progress in Applied Electrical Engineering (PAEE), Koscielisko, Poland, June, 2018.
[23] S. Czapp, Method of earth fault loop impedance measurement without nuisance tripping of RCDs in 3-phase low-voltage circuits, Metrol. Meas. Syst. 2 (26) (2019) 217–227.
[24] S. Czapp, K. Borowski, Verification of safety in low-voltage power systems without nuisance tripping of residual current devices, Electr. Pow. Syst. Res. 172 (July 2019) 260–268.
[25] R. Roskosz, K. Jakubiuk, L. Swędrowski, D. Świsulski, M. Wołoszyk, M. Ziółko, Method and system for measurement of fault loop impedance in low voltage systems with RCDs (in Polish), Polish Patent no. 221693, 2015.
[26] R. Roskosz, E. Musiał, S. Czapp, Method and system for measuring the short circuit loop resistance in the low voltage installations without tripping the ground fault circuit breakers (in Polish), Polish Patent no. 227061, 2017.
[27] KYORITSU, D-LOK function for fault loop measurement without tripping of residual current devices (RCD) in meter KEW4120A. Assessment of fault loop impedance measurement results. Catalogue materials from the KYORITSU company, 2007.

CHAPTER 10

Residual current devices in circuits with surge protective devices

The reliability of supply in electrical installations depends, among others, on the immunity of RCDs to transient leakage currents and surge currents [1–3]. According to IEC 61008-1 [4], RCDs are tested under the surge current waveforms presented in Fig. 10.1.

The first type of test waveform is a waveform 8/20 μs (Fig. 10.1A). The peak value of the test current impulse is equal to 3 kA. General type RCDs may trip during this test; however, some manufacturers offer RCDs that are immune to such a waveform. The technical specification included in [5] informs about accessible Super Resistant A-type RCDs—their surge current withstand capability (8/20 μs) is >3 kA. Standard IEC 61008-1 [4] requires that every S-type (delayed) is able to withstand 3 kA test current.

The second type of test waveform is presented in Fig. 10.1B. The peak value of the waveform is required to be 200 A (for RCDs of $I_{\Delta n} > 10$ mA) [4]. During this test, both general type (no-delayed) and S-type (delayed) RCDs cannot trip.

Selection of RCDs from the point of view of their immunity to the impulse residual current is important in electrical installations equipped with surge protective devices (SPDs). SPDs may be installed on the load side of RCDs (Fig. 10.2A). If so, standard HD 60364-5-534 [6] informs that RCDs may be S-type (delayed) or no-delayed but their immunity to the surge currents shall be at least 3 kA (8/20 μs). In order to avoid unwanted tripping of RCDs, due to their proximity to SPDs, it is possible to install the latter upstream the RCDs (Fig. 10.2B). In such a solution, the surge current flows to the earth upstream the RCD and no residual current occurs. This way of coordination RCDs and SPDs gives higher reliability of supply.

In practice, no-delayed RCDs can have higher immunity to the surge currents than required by the standard [4]. Manufacturers testing RCDs according to this standard may not be aware of the real immunity of their own products beyond the required range of the test. Results of the wider test of RCDs are presented in the report [7]. RCDs from three

Residual Current Devices
https://doi.org/10.1016/B978-0-323-89783-9.00016-3

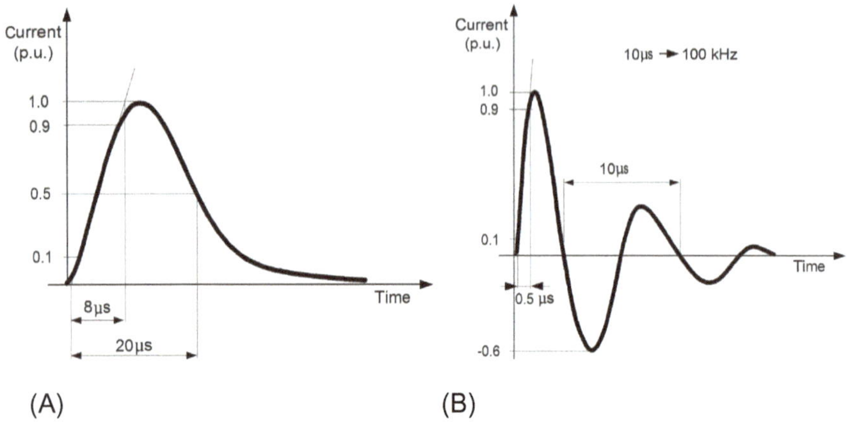

Fig. 10.1 Standard waveforms for testing the immunity of RCDs against surge currents: (A) 8/20 µs, (B) 0.5 µs/100 kHz; according to [4].

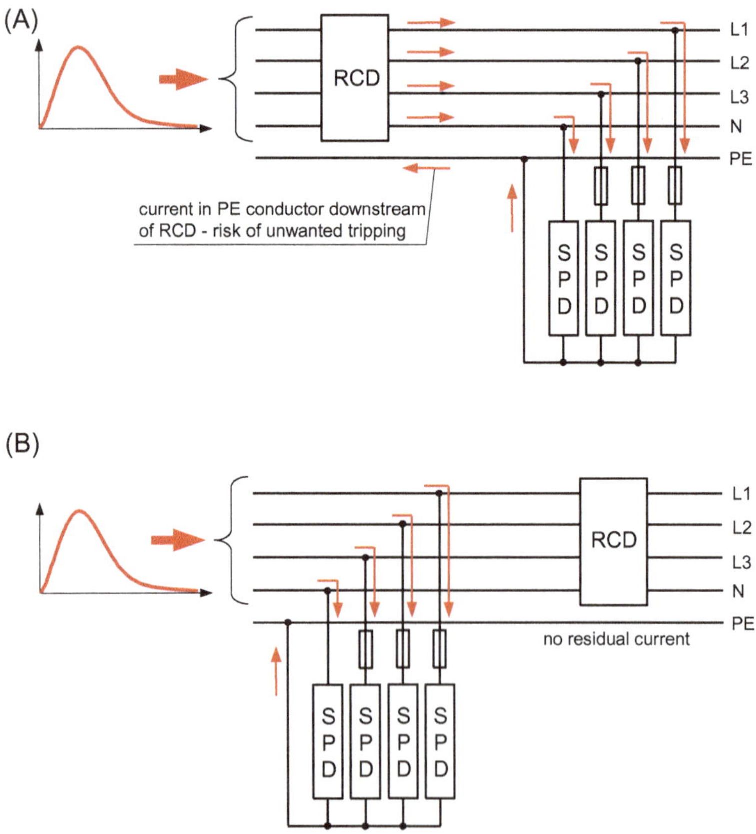

Fig. 10.2 Effect of the surge current on the RCD: (A) risk of unwanted tripping—residual current occurs, (B) no risk of unwanted tripping; SPD—surge protective device.

Table 10.1 Real tripping current (waveform 8/20 μs) of the selected RCDs; according to [7].

$I_{\Delta n}$, I_n	30 mA, 25 A	30 mA, 40 A
Manufacturer 1	5.6 kA	5 kA
Manufacturer 2	3 kA	3 kA
Manufacturer 3	2 kA	2 kA

manufacturers were tested. Their rated residual operating current is equal to $I_{\Delta n} = 30$ mA, but the rated current of the first group is $I_n = 25$ A (three RCDs) and the second group is $I_n = 40$ A (also three RCDs). Results of the test of the immunity to the waveform 8/20 μs, included in the report [7], are presented in Table 10.1. RCDs of manufacturer 1 have the highest immunity. They are able to withstand around 5 kA.

RCDs from manufacturer 2 have immunity around 3 kA—the same level as required for S-type RCDs. The lowest immunity (2 kA) is recorded for RCDs of manufacturer 3.

The above-presented results of the test show that in some cases the behavior of RCDs can be unexpected. Even no-delayed RCDs can withstand the high value of surge currents.

Considerations regarding the location of the RCD relative to SPDs should take into account the type of the low-voltage system (TN or TT). The SPD may get damaged, e.g., a short-circuit inside the SPD may occur, and the type of the low-voltage system then determines the risk of electric shock as well as possible unwanted tripping of the RCD.

In the case presented in Fig. 10.3A, a short-circuit in any SPD connected to the line terminal (here in the line L1) results in the short-circuit current I_k flow in the low-impedance loop, which is typical for TN systems. The current I_k has relatively high value (usually from a few hundred amps to a few kiloamps) and it is high enough to make tripping of the fuse F1. However, the I_k is the residual current for the RCD, so—unfortunately—most likely its tripping will occur as well.

Application of the RCD downstream the SPDs in a TN system (Fig. 10.3B) gives the opportunity to make tripping only by the fuse F1 and protects the RCD against its unwanted tripping. It favors the reliability of supply.

More complicated problems refer to a low-voltage TT system. In a TT system with the RCD installed downstream of the SPDs (what is preferred in TN system), the SPD failure in the line L1 (Fig. 10.4A) does not result in the

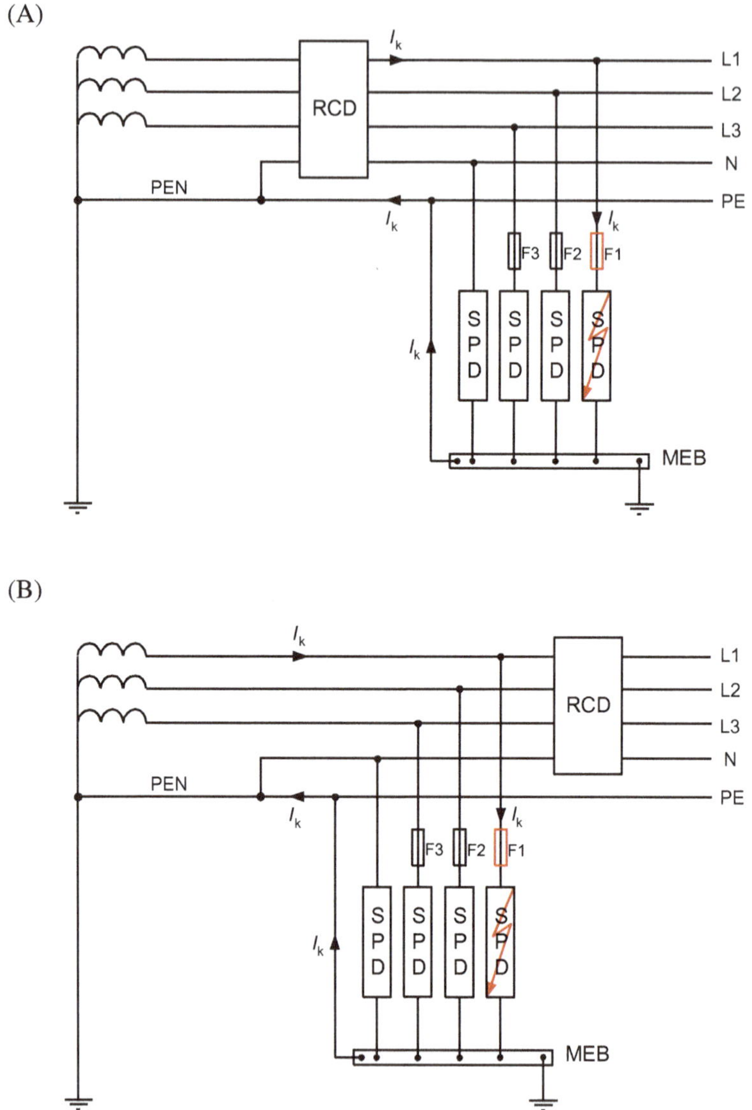

Fig. 10.3 Behavior of the RCD in the case of the surge protective device failure (SPD in the line L1) in a TN system with the 4+0 SPD configuration: (A) RCD upstream the SPDs, automatic disconnection of supply by fuse F1 occurs, the RCD may also trip, (B) RCD downstream the SPDs, automatic disconnection of supply by fuse F1 occurs and no unwanted tripping of the RCD is suspected; I_k—fault current, MEB—main equipotential bonding busbar.

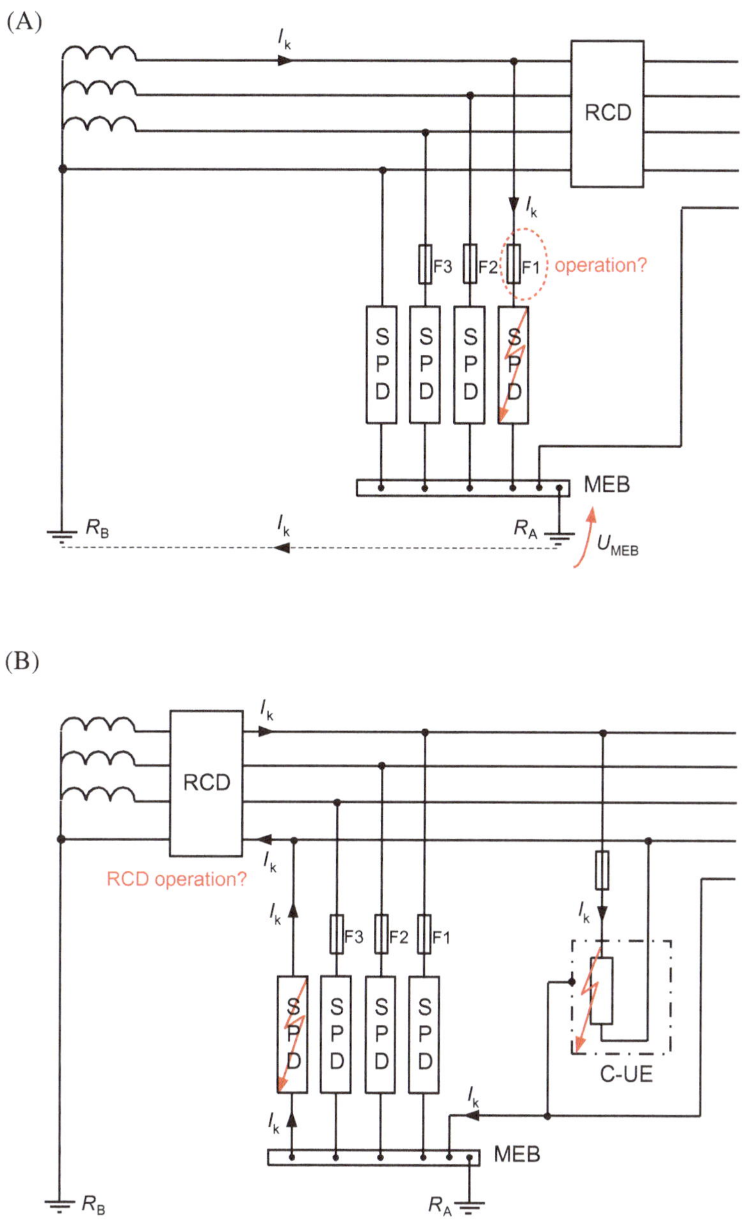

Fig. 10.4 Risk of electric shock in a TT system with the 4+0 SPD configuration: (A) RCD downstream the SPDs, failure of the SPD in the line L1—no automatic disconnection of supply by fuse F1 (high-resistance loop—too low value of the fault current I_k), dangerous voltage U_{MEB} may exist, (B) RCD upstream the SPDs, failure of the SPD in the line N and failure of the current-using equipment—the RCD does not react to the fault current in the current-using equipment; C-UE—current-using equipment, MEB—main equipotential bonding busbar.

operation of the fuse L1. This is due to the high-impedance/resistance loop for the current I_k. This current flows through the earth and it is limited mainly by the following two resistances: R_A and R_B. The sum of these resistances may exceed 10...50Ω, therefore the current I_k is, for example, 10 A (230 V/23 Ω = 10 A). Such a low value of the current I_k can be less than the rated current of the fuse F1 (e.g., $I_{n\text{-}F1} = 25$ A). If so, the current may give, for a long time, dangerous voltage U_{MEB} across the resistance R_A connected to MEB and it introduces to the installation, via the PE conductor, the risk of electric shock.

An alternative solution—the RCD upstream of the SPDs—is not free from the risk of the electric shock and unexpected behavior of the RCD. In the case presented in Fig. 10.4B, it is a short-circuit of the SPD in the neutral terminal N (which may not be noticed immediately) and an insulation fault in the current-using equipment (C-UE). Such two accidents give a relatively low-impedance loop for the I_k. However, the RCD does not detect the insulation fault in the C-UE circuit (no residual current flows), only the fuse in this circuit may react. This is the kind of RCD blinding.

To avoid improper behavior of the RCD in TT system, as well as the risk of electric shock due to SPDs' failure, the SPD configuration called 3+1 should be used (Fig. 10.5). The RCD is installed downstream of the SPDs,

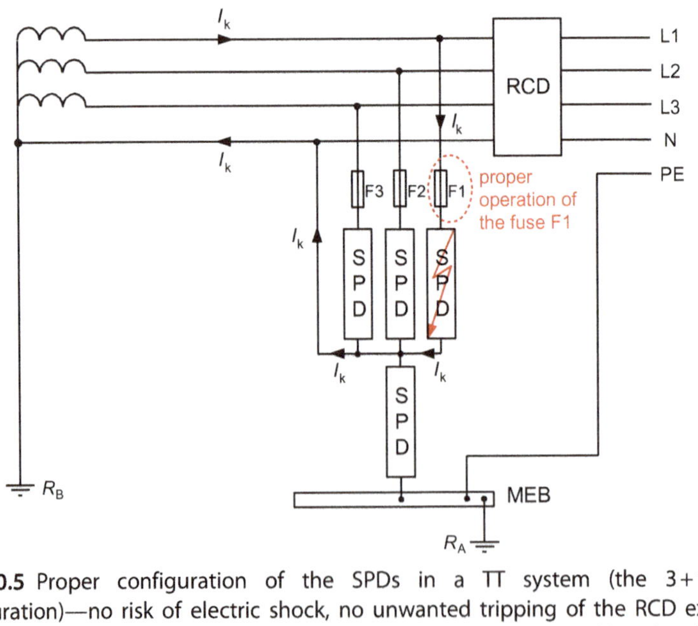

Fig. 10.5 Proper configuration of the SPDs in a TT system (the 3+1 SPD configuration)—no risk of electric shock, no unwanted tripping of the RCD exists in the case of the SPD failure; I_k—fault current, MEB—main equipotential bonding busbar.

which allows to avoid its unwanted tripping. In the case of the SPD failure, the short-circuit current I_k flows in the low-impedance loop L1–N. No RCD tripping and no dangerous voltage across the MEB occur. Moreover, the fuse F1 may disconnect the supply within a short time.

References

[1] C. Roldan-Porta, G. Escriva-Escriva, F.J. Carcel-Carrasco, C. Roldán-Blay, Nuisance tripping of residual current circuit breakers: a practical case, Electr. Pow. Syst. Res. 106 (2014) 180–187.
[2] S. Czapp, Testing of the RCDs immunity to transient residual currents (in Polish), Elektro.info 4 (73) (2009) 48–51.
[3] S. Czapp, K. Borowski, Immunity of residual current devices to the impulse leakage current in circuits with variable speed drives, Elektron. Elektrotech. 8 (19) (2013) 15–18.
[4] IEC 61008-1:2010, Residual current operated circuit-breakers without integral overcurrent protection for household and similar uses (RCCBs) – Part 1: General rules, 2010.
[5] Siemens, Residual Current Protective Devices/Arc Fault Detection Devices. Configuration Manual. Siemens AG, 2014.
[6] HD 60364-5-534:2016, Low-voltage electrical installations – Part 5-53: Selection and erection of electrical equipment – Isolation, switching and control – Clause 534: Devices for protection against transient overvoltages, 2016.
[7] Blitzstromfeste Niederspannungsschaltgeräte, 14209 BR/AiF-Vorhaben-Nr./GAG, Abschlussbericht für den Zeitraum: 01.11.2004 bis 28.02.2007, Technische Universität Ilmenau, Institut für Elektrische Energie- und Steuerungstechnik. Ilmenau, 2007.

CHAPTER 11

Immunity of residual current devices to the impulse leakage currents

Reliability of supply in low-voltage systems depends, among others, on the immunity of the RCDs to the impulse leakage currents which may occur during switching on electronic current-using equipment (a transient state) as well as during normal operation of such equipment. Fig. 11.1A presents an oscillogram of the leakage current (measured in the protective conductor PE) flowing in the circuit supplying only one personal computer. The peak value of this current exceeds 8 A and has a very short decay time. Despite such a short time, the transient state may cause, with a very high probability, unwanted tripping of the 30 mA no-delayed RCD. The cause of the unwanted tripping of the RCD can also be an impulse leakage current, which is met in variable-speed drive circuits [1,2]. Fig. 11.1B depicts a waveform of the leakage current recorded in an example variable-speed drive circuit. There are quasiregular impulses of the frequency which depends on the Pulse Width Modulation frequency. The peak value of the current reaches almost 4 A.

From the early observations of the author, referred to no-delayed RCDs, the conclusion is that theoretically the same RCDs (in terms of the nominal parameters) have various reactions to the impulse currents. To verify the behavior of the RCDs under impulse leakage current, a laboratory test of the 5 selected 30 mA no-delayed RCDs (2 × AC-type and 3 × A-type) has been performed. The structure of the testing circuit as well as an example testing waveform is presented in Fig. 11.2. Such a waveform of controlled parameters was forced as a residual current, and the tripping/no tripping of the particular RCDs were registered. The following parameters of the current were adjusted:

- The peak value: 0.1, 0.5, 1, and 3 A,
- The frequency of the impulses (f): 100, 200, 300, 500, and 1000 Hz.

https://doi.org/10.1016/B978-0-323-89783-9.00001-1

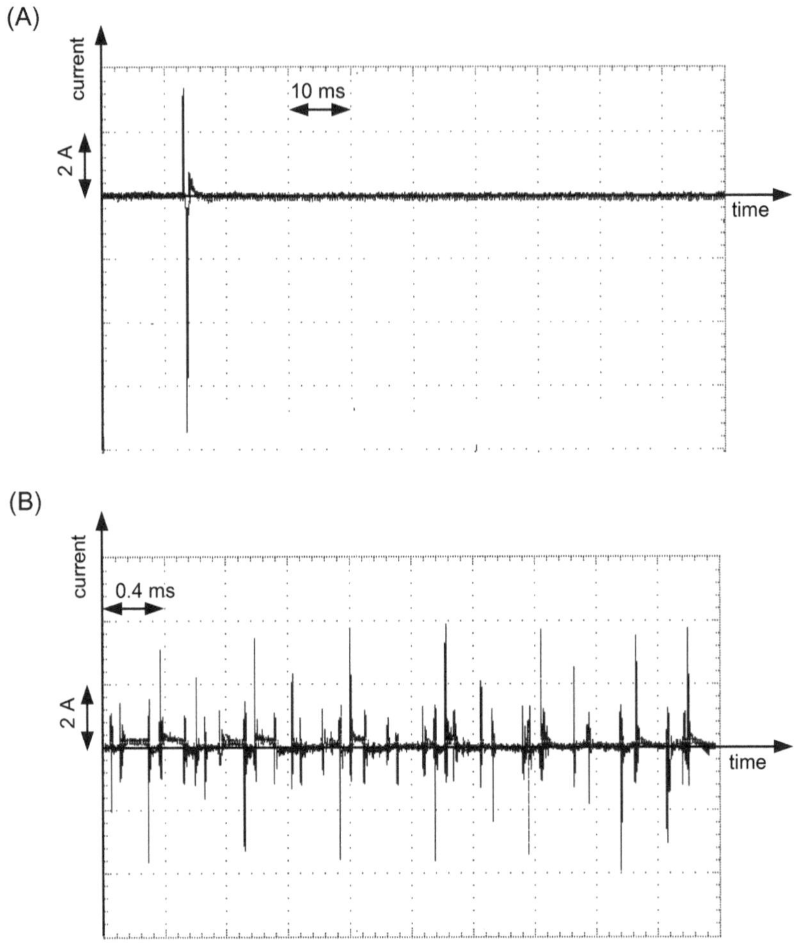

Fig. 11.1 Leakage current measured: (A) during switching on one personal computer, (B) in a variable-speed drive circuit.

Results of the tripping test under impulse current for all 5 RCDs are presented in Fig. 11.3. In the case of the testing current of peak value equal to 0.1 A, tripping of only RCD marked A1i occurred (Fig. 11.3A). Moreover, this tripping was noted only for the frequency of the impulses equal to 100 Hz. Increasing the peak value of the testing current to 0.5 A resulted in the tripping of the RCD no. A1i for frequencies 100 and 200 Hz. Also, RCD marked A2i tripped out but only for frequencies 100, 300, and

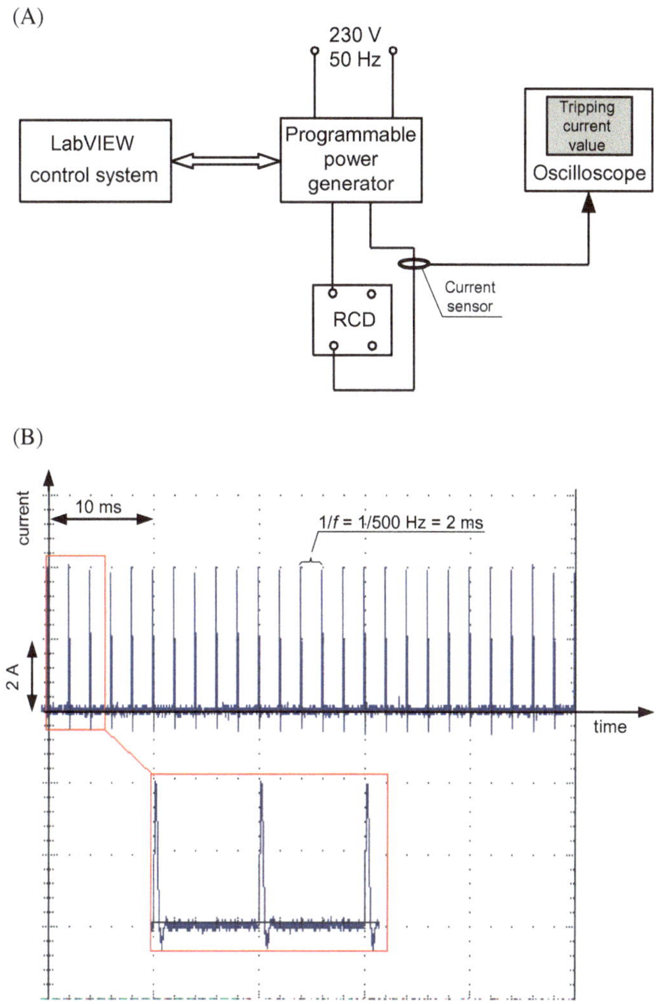

Fig. 11.2 Simplified structure of the testing circuit (A); an example waveform of the testing impulse leakage current and its parameters (B).

500 Hz (Fig. 11.3B). Further increasing the peak value of the current (to 1 A) resulted in the tripping of the following RCDs (Fig. 11.3C):

- A1i—for 100, 200, and 1000 Hz,
- A2i—for all testing frequencies,
- AC1i—for the frequency 100 Hz.

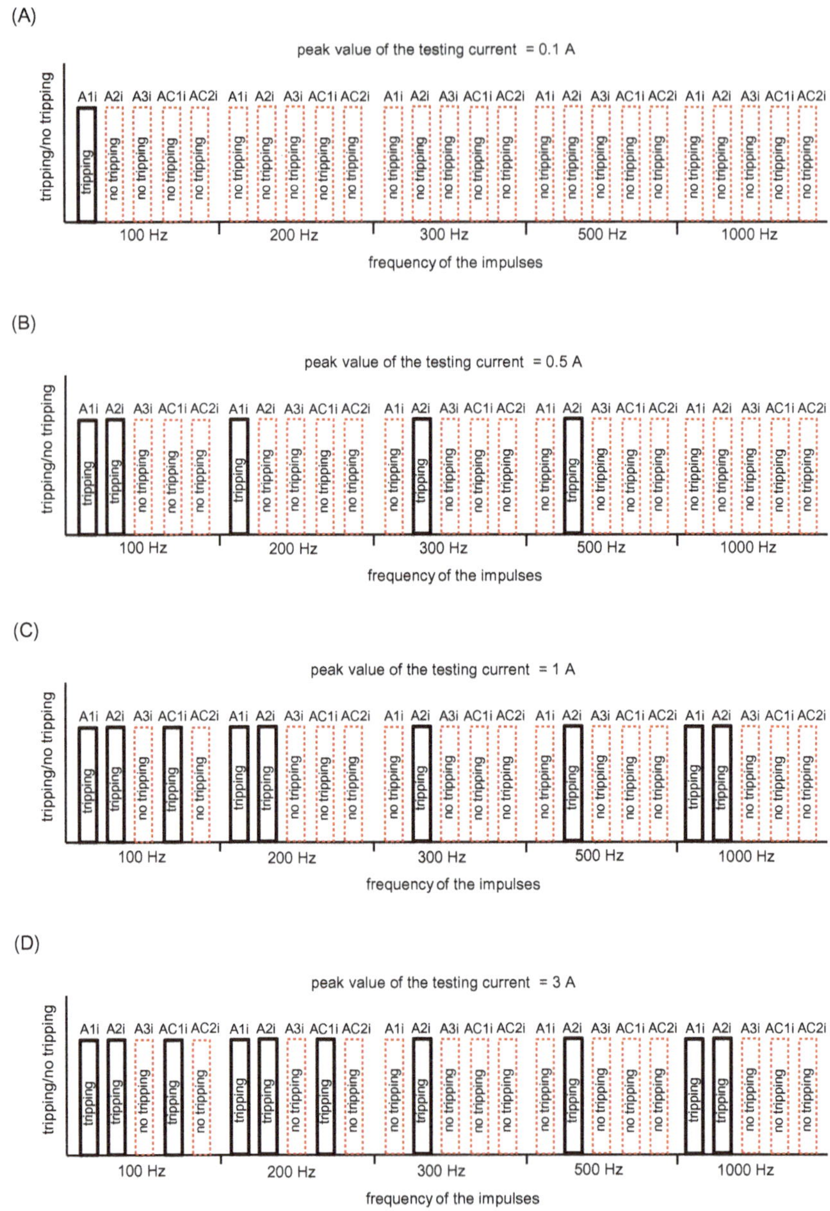

Fig. 11.3 Result of the immunity test of the selected RCDs (AC-type: AC1i and AC2i; A-type: A1i, A2i, and A3i) to the impulse leakage currents. Peak value of the testing current: (A) 0.1 A, (B), 0.5 A, (C) 1 A, (D) 3 A.

The same responses to the testing current were observed for the peak value 3 A but, additionally, the RCD marked AC1i tripped out for frequency 200 Hz (Fig. 11.3D). One can see that RCDs marked A3i and AC2i did not trip at any time during the test.

Taking into account information that all tested RCDs have the same rated residual operating current (30 mA) and that they are no-delayed, there is no possibility to predict their behavior under impulse current in practice, without additional test and analysis.

However, the author's research indicated some correlation between RCDs' behavior under impulse current and RCDs' tripping characteristics as a function of residual current frequency. Analysis of the tripping curves of the RCDs presented in Figs. 11.4 and 11.5 enables to say that RCDs no. A1i (Fig. 11.4A), no. A2i (Fig. 11.4B), and no. AC1i (Fig. 11.5A) trip out within the relatively wide frequency range—they react to the testing current within the assumed range of 50–1000 Hz. Therefore one can observe that their favorable frequency tripping curves translate their sensitivity to the impulse leakage current. Obviously, the lower frequency of the impulse current, the more probability of the tripping of the RCD. On the other hand, RCDs no. A3i (Fig. 11.4C) and no. AC2i (Fig. 11.5B) detect only low-frequency residual currents (practically within the range 50–350 Hz) and it results in their no reaction to the impulse currents, even of peak value equal to 3 A. Paradoxically, RCDs having bad tripping characteristic as a function of frequency (and therefore not suitable to the circuit of high/mixed-frequency residual currents) are immune to the transient/impulse leakage current (favorable effect). In this place, it is worth mentioning that frequency-dependent tripping curves (at least up to 1000 Hz) of any RCD type are recommended to be presented by manufacturers. Such curves would be very useful during the selection of the RCDs (especially of the no-delayed type) to circuits with electronic loads/power electronics.

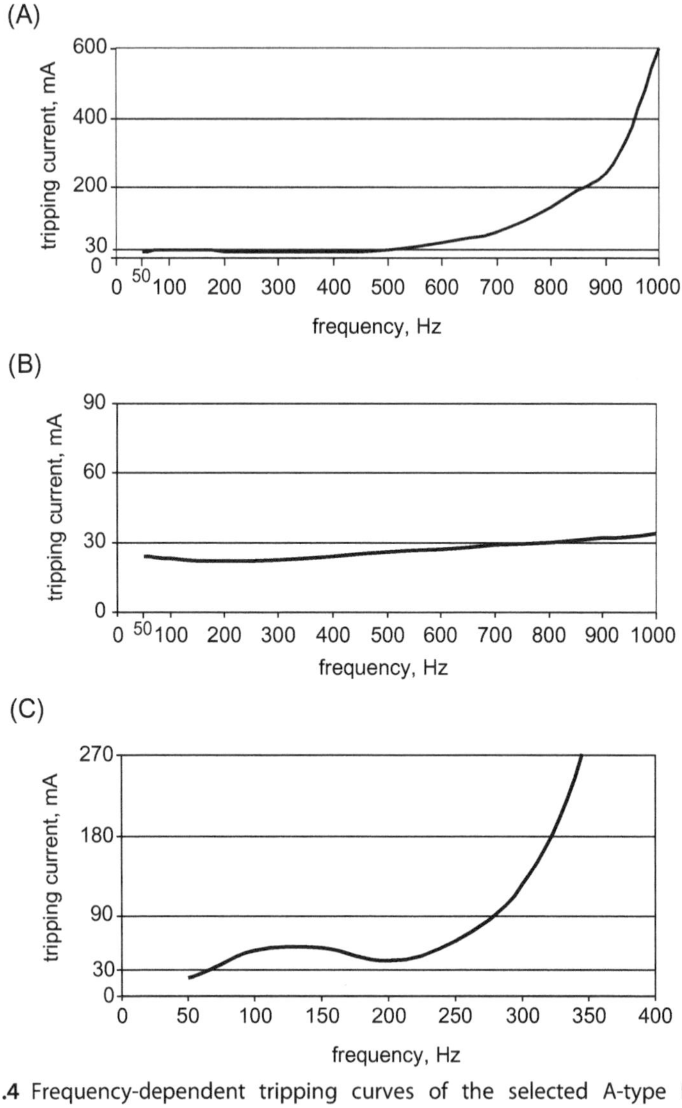

Fig. 11.4 Frequency-dependent tripping curves of the selected A-type RCDs of $I_{\Delta n} = 30\,\text{mA}$: (A) RCD no. A1i, (B) RCD no. A2i, (C) RCD no. A3i.

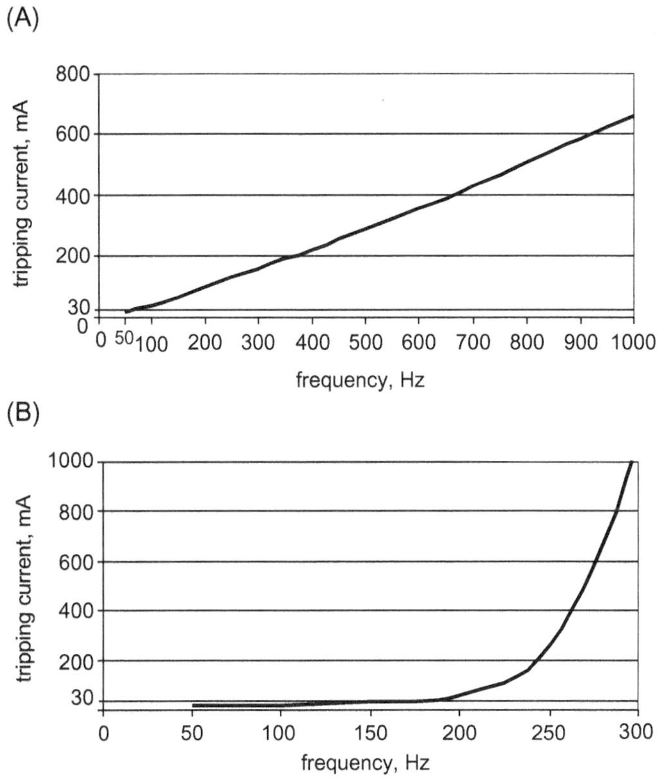

Fig. 11.5 Frequency-dependent tripping curves of the selected AC-type RCDs of $I_{\Delta n} = 30\,\text{mA}$: (A) RCD no. AC1i, (B) RCD no. AC2i.

References

[1] J. Schoneck, Y. Nebon, LV protection devices and variable speed drives, Cahier technique no. 204. Schneider Electric, 2002.

[2] S. Czapp, J. Guzinski, Electric shock hazard in circuits with variable-speed drives, Bull. Pol. Ac.: Tech. 3 (66) (2018) 361–372.

CHAPTER 12

Verification of residual current devices

12.1 General information

Residual current devices are more susceptible to damage than overcurrent protection devices (fuses or circuit-breakers). For this reason, it is necessary to verify their operation during the initial and periodic verification of the electrical installations. The following tests should be performed:

- simple tripping test by pushing TEST button,
- verification of the tripping current,
- verification of the break time.

The first from the aforementioned tests (pushing the TEST button) is very simple, no professional tools are necessary and can be done even by a non-skilled person. This test verifies the RCD tripping but only roughly. In households, it is recommended to do this test once a month (at least once every 6 months). However, RCDs installed in Polish construction or demolition sites are obliged to be verified by TEST button before every beginning of work (practically every day)—this requirement is specified by the Regulation of the Minister of Infrastructure of February 6, 2003 on occupational health and safety during construction works (Journal of Laws, 2003, no. 47, item 401).

The other two aforementioned tests are more advanced and are commented on in separate sections.

12.2 Verification of the tripping current

It is required that the residual current device has the real tripping current between rated residual nonoperating current $I_{\Delta no}$ and rated residual operating current $I_{\Delta n}$. The verification of the RCD from the point of view of this criterion is especially important in the case of a new installed RCD. The RCD could have been damaged during the transportation or installation process. In order to check whether the real tripping current of the RCD complies with the normative tripping characteristics

Residual Current Devices
https://doi.org/10.1016/B978-0-323-89783-9.00007-2

presented in Fig. 3.11, the following procedure, according to the standard [1], is recommended to be performed. The testing residual current is steadily increased from a value not higher than $0.2I_{\Delta n}$ to reach value $1.0I_{\Delta n}$ within 30 s. The tested RCD should not trip for currents up to rated residual nonoperating current $I_{\Delta no}$ but should trip for currents not higher than the rated residual operating current $I_{\Delta n}$. The test is required at least for the sinusoidal waveform and then the real tripping current $I_{\Delta r}$ should be within the range: $0.5I_{\Delta n} < I_{\Delta r} \leq 1.0I_{\Delta n}$. The connection of the testing instrument to the circuit with the RCD is presented in Fig. 12.1.

For the performance of the test, the simplest laboratory tools can be employed (Fig. 12.1A): a resistor of variable resistance and an ammeter for sinusoidal currents. Decreasing this resistance, the residual/test current rises and the tripping threshold of the RCD should be read from the ammeter. In practice, the professional instrument for RCDs testing is used. The RCD tester is connected to the circuit as presented in Fig. 12.1B. The entire measurement is automatic and the result is displayed by the tester.

Verification of the real tripping current of the 14 selected RCDs from 8 manufacturers has been performed in laboratory conditions. The following 30 mA RCDs were selected for the test:

- 5 × AC-type (marked AC1, AC2, ...AC5),
- 5 × A-type (marked A1, A2, ...A5),
- 2 × F-type (marked F1, F2),
- 2 × B-type (marked B1, B2).

The most important parameters of the tested RCDs, as well as the results of the test of the real tripping current, are presented in Table 12.1. The recorded tripping current varies from 16.1 mA (RCDs: AC1 and A3) to 29.1 mA (RCD AC4). Thus all tested RCDs have the tripping current within the normative range—higher than $0.5I_{\Delta n}$ (15 mA) but not higher than $1.0I_{\Delta n}$ (30 mA). Most of the tested RCDs have real tripping current close to the geometric mean from the aforementioned values:

$$I_{\Delta r} = \sqrt{(0.5I_{\Delta n}) \cdot (1.0I_{\Delta n})} = 0.71I_{\Delta n} = 0.71 \cdot 30\,\text{mA} = 21.3\,\text{mA} \quad (12.1)$$

Real tripping current close to the geometric mean value is very convenient from the reliability of supply and safety points of view (relatively high margin from the limit values is maintained).

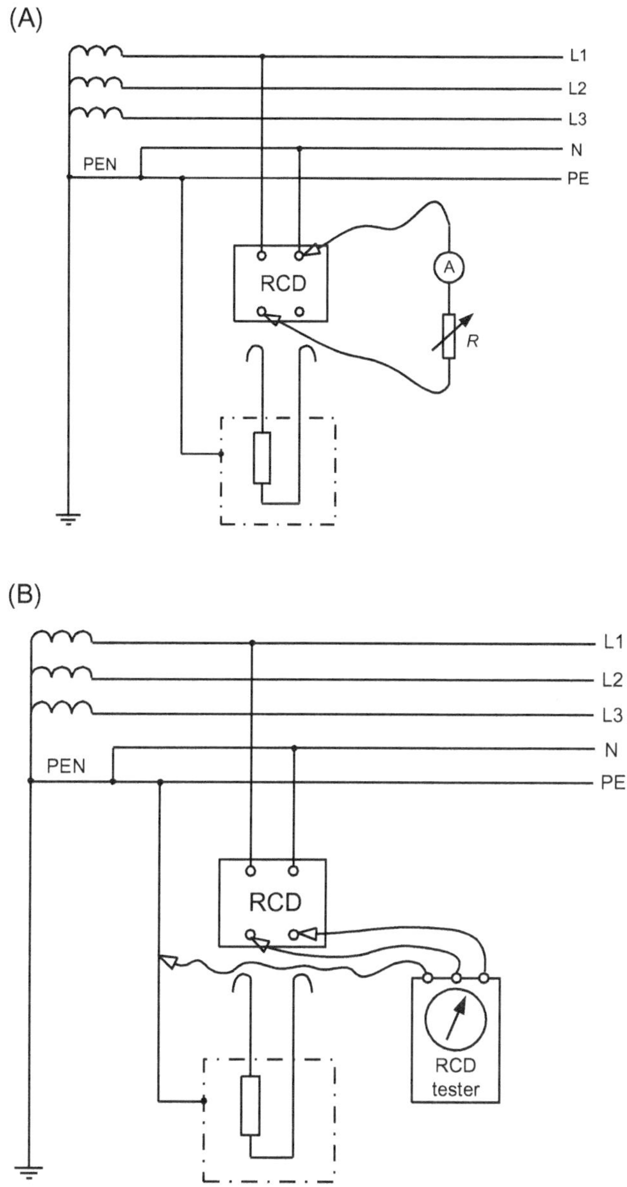

Fig. 12.1 Connection of the testing instrument during the verification of the real tripping current of the RCD: (A) the use of the simple laboratory elements (ammeter, resistor *R*), (B) the use of the professional RCD tester.

Table 12.1 Measured tripping current (for sinusoidal waveform) of the selected 14 RCDs of $I_{\Delta n} = 30\,mA$.

RCD type	RCD symbol	Rated current (A)	Number of poles (–)	Real tripping current (mA)	Manufacturer
AC	AC1	25	2	16.1	Man_4
	AC2	25	2	20.4	Man_5
	AC3	16	2	24.8	Man_1
	AC4	16	2	29.1	Man_8
	AC5	40	4	21.9	Man_6
A	A1	25	2	21.9	Man_4
	A2	25	4	23.3	Man_5
	A3	6	2	16.1	Man_8
	A4	25	2	26.2	Man_8
	A5	25	2	20.4	Man_7
F	F1	40	2	20.4	Man_1
	F2	40	2	21.9	Man_2
B	B1	63	4	21.9	Man_3
	B2	40	4	23.3	Man_1

Careful examination of the drawing (Fig. 12.1) allows to conclude that the protected final circuit is disconnected from the output terminals of the RCD. This is deliberate, and in some cases, necessary. Some types of current-using equipment, especially electronic receivers, are characterized by a relatively high leakage current. The leakage current can be resistive or capacitive and interacts with the current which is forced by the RCD tester. The effect of the leakage current on the tripping threshold of the RCDs, during the verification of their real tripping current, has been checked in laboratory conditions. In a three-phase final circuit, a four-pole RCD of a rated residual operating current 30 mA was installed. In the first stage of the test, a capacitive leakage current of value equal to 5 mA was forced in phase L1. The arrangement of the circuit during the forcing of the leakage current and the RCD testing is depicted in Fig. 12.2A.

The leakage current I_{leak} adds itself geometrically to the testing current I_{test}. In consequence, the value displayed by the tester isn't the proper one referred to the tested RCD. Table 12.2 presents the results of the test when the leakage current was forced. Measurement of the tripping current between L1 and PE gives an average value equal to 20.85 mA. Similar measurement between L2 and PE gives a higher value (23.85 mA) whereas between L3 and PE gives only 15.9 mA (close to the lower limit equal to

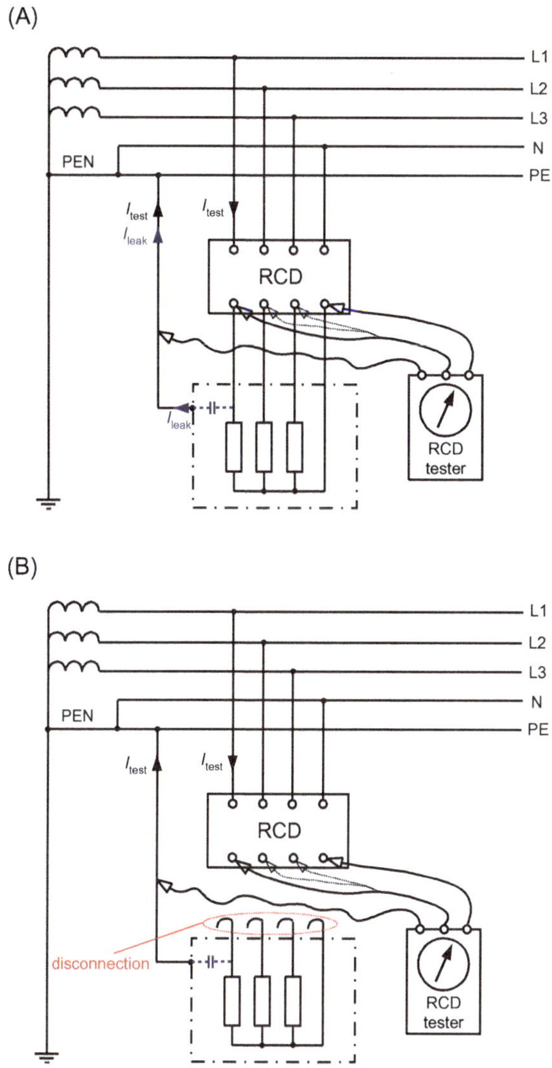

Fig. 12.2 Connection of the testing instrument during the verification of the real tripping current of the RCD in a three-phase circuit: (A) final circuit is connected to the RCD, (B) final circuit is not connected to the RCD.

15 mA). The dispersion of values is relatively high and one can conclude that the internal release circuit of the tested RCD does not operate properly. Moreover, in the case of an even higher value of the leakage current, the result of the measurement L3–PE could give a value less than 15 mA, which qualifies the RCD as damaged.

Table 12.2 Results of the RCD ($I_{\Delta n}=30\,mA$) testing in circuit with the capacitive leakage current L1–PE (5 mA).

Testing between	Tripping current indicated by tester (I_{test}) (mA)		
	Trial 1	**Trial 2**	**Average value**
L1–PE	21.0	20.7	20.85
L2–PE	23.7	24.0	23.85
L3–PE	15.6	16.2	15.9

The explanation of the leakage current impact on the tripping current obtained during the aforementioned measurements is presented in the form of the phasor diagrams in Fig. 12.3. When the tester is connected to the phase L1, the capacitive leakage current I_{leak} is orthogonal to the resistive testing current I_{test} forced and displayed by the tester as the tripping current (Fig. 12.3A). The geometric sum of the above-mentioned currents gives current I_r, which is the real tripping current of the tested RCD. In such a case, the value of the current indicated by the tester I_{test} (average 20.85 mA—see Table 12.2) is slightly lower than the real tripping current I_r without the leakage current (average 21.3 mA—see Table 12.3). The relation between the phasors during the test in phase L2 is presented in Fig. 12.3B. The testing current I_{test} displayed by the tester is higher than the real tripping current I_r. The measured values are as follows: $I_{test}=23.85\,mA$ (average value in Table 12.2), $I_r=21.0\,mA$ (average value in Table 12.3). The highest difference between the current displayed by the tester I_{test} and the tripping current I_r without the leakage current I_{leak} occurs during the measurement in phase L3 (Fig. 12.3C): $I_{test}=15.9\,mA$ (Table 12.2), $I_r=20.85\,mA$ (Table 12.3). The resulting value is almost the arithmetic sum of the values $I_{test}+I_{leak}\cong 15.9\,mA+5\,mA=20.9\,mA$.

Note that in the case of a leakage current flowing in other phases as well as in the case of a resistive leakage current, even of the same value (5 mA), the results of the testing of the same RCD will be different. To avoid incorrect evaluation of the operation of the tested RCD, the final circuit of the electrical installation should be disconnected from the output terminals of the tested RCD, as presented in Fig. 12.2B. This ensures that any possible leakage currents do not flow and do not affect the results of the measured tripping current of the RCD. Results of such a proper measurement are summarized in Table 12.3.

It is worth adding that standard IEC 61557-6 [2] mentions also the verification of the RCD under the testing current equal to $0.5I_{\Delta n}$. This testing

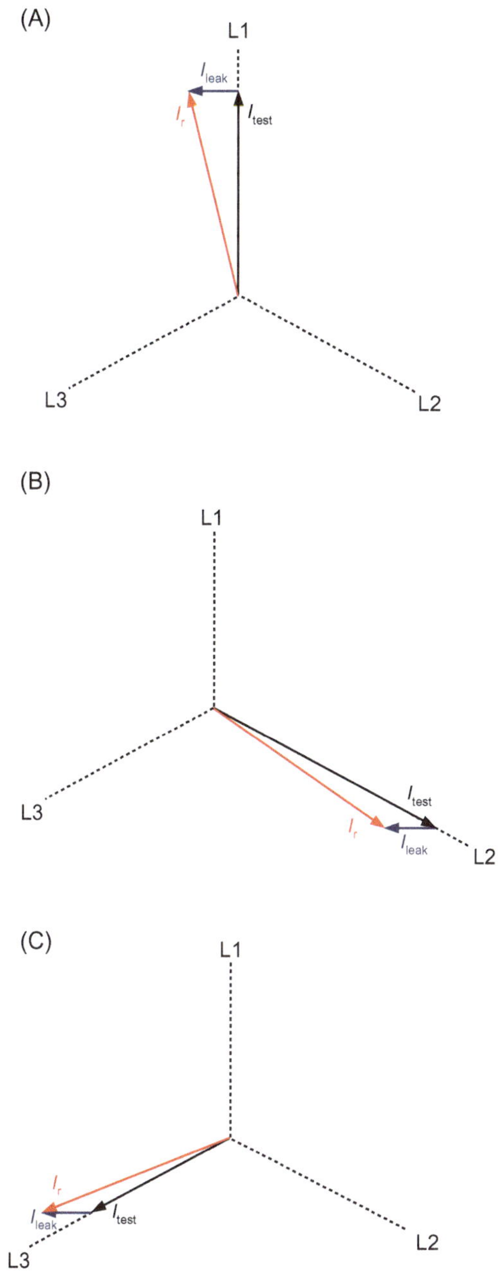

Fig. 12.3 Phasor diagrams of the currents flowing during the RCD ($I_{\Delta n} = 30\,mA$) testing in presence of the capacitive leakage current: (A) tester connected to the phase L1, (B) tester connected to the phase L2, (C) tester connected to the phase L3; I_{leak}—capacitive leakage current (5 mA) flowing from phase L1, I_{test}—current displayed by the RCD tester, I_r—real tripping current of the RCD without the leakage current.

Table 12.3 Results of the RCD ($I_{\Delta n} = 30\,mA$) testing in a circuit without the capacitive leakage current.

Testing between	Tripping current indicated by tester ($I_{test} = I_r$) (mA)		
	Trial 1	Trial 2	Average value
L1–PE	21.3	21.3	21.3
L2–PE	21.0	21.0	21.0
L3–PE	20.7	21.0	20.85

current should be forced suddenly for a period of at least 0.2s. During this test, the RCD is not allowed to trip.

The verification of the tripping current "in situ" should be conducted at least with the use of the sinusoidal testing current. However, standard IEC 61557-6 [2] specifies the use of other testing waveforms if A-type or B-type RCDs are checked. According to this standard, for A-type RCDs the additional recommended waveform is a pulsating direct current (single-pulse DC). B-type RCDs are required to be tested also under smooth direct current. For testing currents of other waveforms, the permissible tripping range is wider than for the 50Hz sinusoidal waveform. If one wants to perform a detailed test of RCDs, to verify their tripping current according to RCDs' product standards [1,3,4], the normative tripping ranges of RCDs, for various waveforms, are applicable as in Table 12.4.

12.3 Verification of the break time

The means of protection in the case of fault—automatic disconnection of supply—is effective provided that a disconnecting device opens the circuit within the required time, indicated by the standard [5]. In electrical installations of the nominal voltage 230/400V, for socket-outlet circuits up to 63 A and for final circuits supplying only fixed connected current-using equipment up to 32 A, the aforementioned standard requires that the disconnection time is not higher than:

- 0.4s in TN systems,
- 0.2s in TT systems.

An RCD may be used as a disconnecting device. Time–current tripping curves of RCDs are depicted in Fig. 3.11. In the case of the residual current flow equal to a rated residual operating current ($I_{\Delta n}$) of RCDs, their tripping is ensured within the time not exceeding 0.3s. Therefore, in TN systems, to fulfill

Table 12.4 Permissible tripping ranges of RCDs for various waveforms.

Testing waveform	Lower limit	Upper limit		Tested type of RCDs	Comments
		$I_{\Delta n} < 30\,mA$	$I_{\Delta n} \geq 30\,mA$		
Sinusoidal 50/60 Hz	$0.5I_{\Delta n}$	$1.0I_{\Delta n}$		AC, A, F, B	
Fulsating DC:					
– Without phase-angle control (half-wave); delay angle: 0°	$0.35I_{\Delta n}$	$2.0I_{\Delta n}$	$1.4I_{\Delta n}$	A, F, B	
– With phase-angle control; delay angle: 90°	$0.25I_{\Delta n}$	$2.0I_{\Delta n}$	$1.4I_{\Delta n}$		
135°	$0.11I_{\Delta n}$	$2.0I_{\Delta n}$	$1.4I_{\Delta n}$		
Half-wave DC (delay angle 0°) + smooth DC component 6 mA (or 10 mA in the case of the F-type RCD)	–	$2.0I_{\Delta n}$	$1.4I_{\Delta n}$	A, (F)	The upper limit is the rms value due to the half-wave DC
Half-wave DC + smooth DC component 10 mA (or 0.4 times the rated residual operating current, whichever is the highest value)	–	$2.0I_{\Delta n}$	$1.4I_{\Delta n}$	F, B	The upper limit is the rms value due to the half-wave DC
Mixed frequency AC current (as presented in Table 3.5)	$0.5I_{\Delta n}$	$1.4I_{\Delta n}$		F, B	
Sinusoidal AC 50 Hz + smooth DC component 10 mA (or 0.4 times the rated residual operating current, whichever is the highest value)	–	$1.0I_{\Delta n}$		B	The tripping current is the rms value due to the AC 50 Hz
DC waveform from rectifying circuits supplied from two phases	$0.5I_{\Delta n}$	$2.0I_{\Delta n}$		B	

Continued

Table 12.4 Permissible tripping ranges of RCDs for various waveforms—cont'd

Testing waveform	Lower limit	Upper limit		Tested type of RCDs	Comments
		$I_{\Delta n} < 30\,mA$	$I_{\Delta n} \geq 30\,mA$		
DC waveform from rectifying circuits supplied from three phases	$0.5I_{\Delta n}$	$2.0I_{\Delta n}$		B	Does not apply to 2-pole B-type RCDs
Smooth direct current	$0.5I_{\Delta n}$	$2.0I_{\Delta n}$		B	
Sinusoidal of higher frequency:				B	
150 Hz	$0.5I_{\Delta n}$	$2.4I_{\Delta n}$			
400 Hz	$0.5I_{\Delta n}$	$6I_{\Delta n}$			
1000 Hz	$1.0I_{\Delta n}$	$14I_{\Delta n}$			

requirements of standard HD 60364-4-41 (0.4 s), it is sufficient that the earth fault current reaches the value of the rated residual operating current of the RCD. It is different in the case of the TT systems. To ensure disconnection of supply within 0.2 s, the RCD is required to achieve the residual current flow equal to at least $2I_{\Delta n}$ (the tripping within 0.15 s). Thus, to reach effective protection by automatic disconnection of supply, it is important that RCDs keep the break time within the normative range. Moreover, in the case of the 30 mA RCD, its time–current tripping curve is favorable in terms of preventing ventricular fibrillation. Fig. 12.4 presents a comparison of the time–current tripping curve of the 30 mA RCD with the threshold of ventricular fibrillation. The time–current tripping curve is situated in such a way that it does not cross the trace c_1 presenting the conventional threshold of ventricular fibrillation.

The real break times of the 14 RCDs mentioned in the previous Section 12.2 have been verified in laboratory conditions. During the test, the sinusoidal testing current was forced by a typical meter accessible on the market. Such a meter enables for forcing the testing current of values $I_{\Delta n}$, $2I_{\Delta n}$, and $5I_{\Delta n}$, for which the standard [1] indicates the maximum permissible break times. In the case of the 30 mA RCD, the testing current is, respectively, 30 mA, 60 mA, and 150 mA.

Results of the test are depicted in Fig. 12.5. The values displayed in the form of bars indicate the maximum break time from a few trials performed for each RCD.

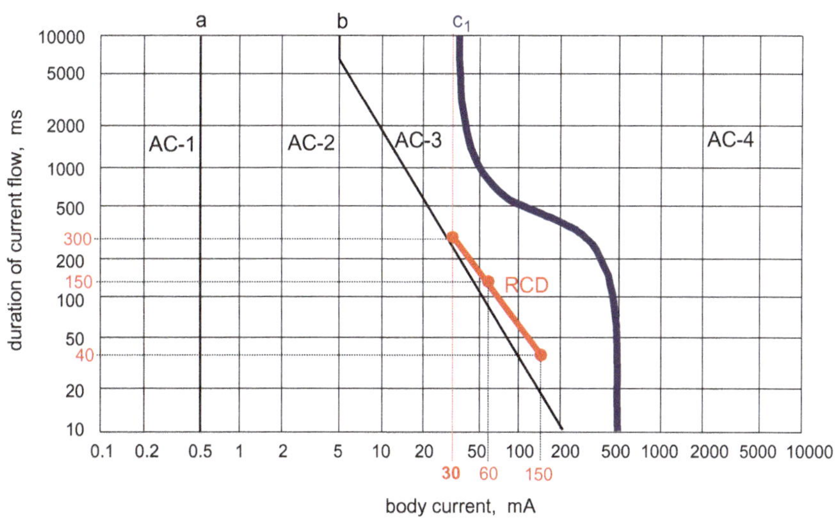

Fig. 12.4 A comparison of the normative time–current tripping curve of the 30 mA RCD (red trace) with the conventional threshold of ventricular fibrillation c_1 (blue trace); areas AC 1, AC-2, AC-3, AC-4 are described in Section 4.2.

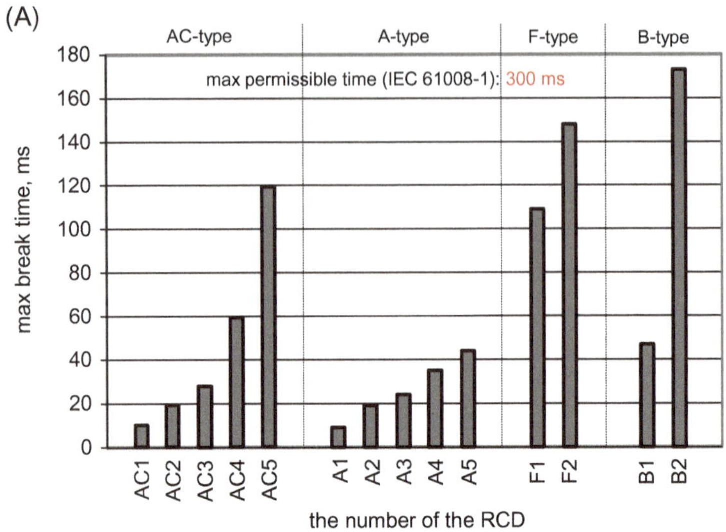

Fig. 12.5 Maximum break time (measured) of the selected 14 RCDs ($I_{\Delta n} = 30\,mA$) under the testing current: (A) 30 mA,

(Continued)

For a given value of the testing current, the measured break times are clearly varied. For example, if one compares RCDs AC1 vs. B2 for the testing current 30 mA (Fig. 12.5A), the former has the break time around 10 ms, whereas RCD B2 has the time close to 180 ms. What is more, a comparison of the break times for the same type of RCDs (e.g., AC-type) allows one to notice that these times are also significantly varied: RCD AC1 has around 10 ms but RCD AC5 has around 120 ms (Fig. 12.5A). If one compares bars from Fig. 12.5C, RCD AC1 has a time less than 10 ms but RCD AC5 has a time over three times more. The break time of the particular RCD depends on the magnetic properties of the current transformer, the type of the electromechanical relay (compare Fig. 2.1 with Fig. 2.3), and any additional components installed in the secondary circuit of the RCD's current transformer. Moreover, B-type RCDs have a very advanced system for detection and identification of the residual current, which is assisted by auxiliary voltage derived from the power network. All these aspects influence the tripping time of RCDs. In the case of the tested RCDs, every device has reached break times lower than the maximum permissible by the standard [1]—all RCDs fulfill the normative requirements. Therefore different and even very long break times should not be a cause for concern when testing RCDs in practice, unless the permissible normative value is exceeded.

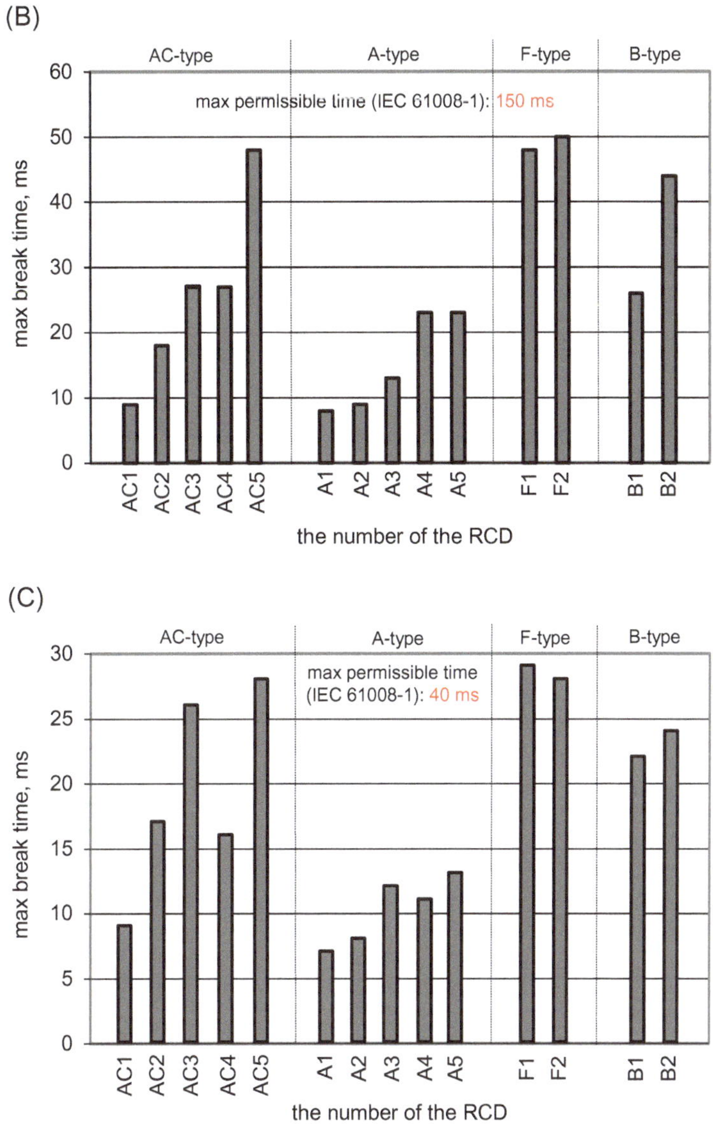

Fig.12.5, cont'd (B) 60 mA, (C) 150 mA.

12.4 Fault diagnosis in circuits with residual current devices

When testing the system of protection against electric shock equipped with RCDs, one can notice their improper operation. The cause of such an operation may exist in the RCD or in the protected circuits. Table 12.5 presents

Table 12.5 Possible faults in circuits with RCDs as well as methods for their detection and elimination.

Symptoms in the circuit	Cause	Diagnosis	Elimination
RCD does not trip under pushing the TEST button.	The main contacts are welded.	It is not possible to open the circuit using the main handheld "0–1" switch.	Replace the RCD with a new one.
	The RCD is not properly connected, e.g., a four-pole RCD is installed in a single-phase circuit but the poles with TEST button circuit are not energized.	Verification of the voltage presence in consecutive poles.	Connection of the single-phase circuit to the poles energizing TEST button circuit.
	Internal TEST circuit is broken due to the failure of the resistor or internal connections.	Force a current exceeding $I_{\Delta n}$, through only one pole. The RCD shall trip.	Replace the RCD with a new one.
	Electromechanical relay is damaged or secondary circuit of the RCD's current transformer has no continuity.	Force a current exceeding $I_{\Delta n}$, through only one pole. The RCD does not trip.	Replace the RCD with a new one.
Dangerous touch voltage is present on the exposed-conductive-parts in the protected circuit but the RCD does not trip.	Failure of the RCD.	No tripping of the RCD either under pushing the TEST button or forcing a current exceeding $I_{\Delta n}$, through only one pole.	Replace the RCD with a new one.
	Improper connection of the conductors to the RCD: PE or PEN conductor goes through the RCD.	Visual inspection of the conductors' connections with the RCD.	Improve the connections.

	PE conductor is connected with N conductor downstream the RCD.	Visual inspection of the connection in the protected circuit.	Improve the connections.
	No PE conductor is installed in the circuit or PE conductor has no continuity.	Verify the presence or the continuity of the PE conductor.	Replace the conductor (conductors) with a proper one.
	In TT system—too high earth electrode resistance.	Measure the earth electrode resistance.	Expand the earthing installation to obtain permissible value of the earth electrode resistance.
	Touch voltage is carried via PE conductor from another circuit.	Open the RCD. The touch voltage still exists on the exposed-conductive-parts.	Verify other circuits supplied from the same distribution board (also the circuit supplied this distribution board).
	Touch voltage is the effect of the residual current waveform shape not detected by the installed RCD (e.g., AC-type RCD is in the circuit generating a pulsating DC waveform).	Visual inspection of the current-using equipment installed in the protected circuit (e.g., rectifiers are installed). Verify the type of the installed RCD.	Replace the RCD with a proper type (e.g., A-type instead of AC-type).

Continued

Table 12.5 Possible faults in circuits with RCDs as well as methods for their detection and elimination—cont'd

Symptoms in the circuit	Cause	Diagnosis	Elimination
Frequent nuisance tripping of the RCD during the switching-on the protected circuit (or closing the RCD).	PE conductor is connected with N conductor downstream of the RCD.	Visual inspection of the connection in the protected circuit.	Improve the connections.
	Failure of the RCD (short-circuit in the TEST circuit, damage of the secondary circuit of RCD's current transformer, failure of the electromechanical relay).	Disconnect the protected circuit from the output terminals of the RCD. If the problem still exists, the RCD has to be replaced.	Replace the RCD with a new one.
	Relatively high value of the transient or steady-state leakage current in the protected circuit.	Disconnect the protected circuit from the output terminals of the RCD. The nuisance tripping should vanish.	Install a short-time delayed RCD (remedy for transient leakage currents). Install an RCD of the higher $I_{\Delta n}$, if permissible (remedy for high-value steady-state leakage currents).
	Neutral conductor of the protected circuit (downstream of the RCD) is connected with a neutral conductor of another circuit. In consequence, load currents flowing in these neutral conductors are mixed.	Visual inspection of the connection in the neutral conductor in the protected circuit.	Disconnect the neutral conductor downstream of the RCD from any other neutral conductors.

the most common faults in circuit with RCDs as well as methods for their detection and elimination.

References

[1] IEC 61008-1:2010, Residual current operated circuit-breakers without integral overcurrent protection for household and similar uses (RCCBs)—Part 1: General rules, 2010.
[2] IEC 61557-6:2007, Electrical safety in low voltage distribution systems up to 1 000 V a.c. and 1 500 V d.c.—Equipment for testing, measuring or monitoring of protective measures—Part 6: Effectiveness of residual current devices (RCD) in TT, TN and IT systems, 2007.
[3] IEC 60755:2017, General safety requirements for residual current operated protective devices, 2017.
[4] IEC 62423:2009, Type F and type B residual current operated circuit-breakers with and without integral overcurrent protection for household and similar uses, 2009.
[5] HD 60364-4-41:2017, Low-voltage electrical installations—Part 4-41: Protection for safety—Protection against electric shock, 2017.

CHAPTER 13
Summary

Residual current devices are integral elements of the modern low-voltage electrical installations. Undoubtedly, they contribute to the safety of people utilizing low-voltage electrical equipment. They are the last safeguard in case of the carelessness of the users, especially in case of direct contact with live conductors.

An analysis of the evolution of the IEC/EN standards enables to say that the obligatory use of RCDs is being increased from version to version of these standards. This tendency will likely continue—mainly regarding RCDs of a rated residual operating current not exceeding 30 mA. Such RCDs will be installed in most low-voltage circuits.

Increasingly, new types of electric receivers, especially those powered via power electronic converters, are putting a challenge on the RCDs' constructors. The RCDs have to detect more and more complicated shapes of residual currents produced by such converters/circuits and, simultaneously, should not be activated by the natural leakage currents. In the nearest perspective, it is expected that RCDs will have to detect very high-frequency components of mixed-frequency earth fault currents. The requirement for the detection of more complicated shapes of the residual currents forces more advanced constructions of RCDs. These advanced constructions cannot influence the worsened reliability of their operation. Thus high reliability of the RCD operation is the next challenge for RCDs' manufacturers.

Residual Current Devices
https://doi.org/10.1016/B978-0-323-89783-9.00017-5

Appendix A

Examples of RCD selection

Example 1

The electrical installation presented in Fig. A.1 has three 1-phase final circuits supplied from different phases (L1, L2, and L3). In each circuit, a resistive natural leakage current of value $I_R = 8\,\text{mA}$ flows. Select a rated residual operating current $I_{\Delta n}$ of the RCD to be installed in the 3-phase main circuit.

Calculations and results

In the above-presented installation, there are three 1-phase circuits that can be switched on independently. To select an RCD, it is necessary to evaluate such a configuration of the operation of the final circuits, which gives the maximal leakage current seen from the point of the RCD location. Fig. A.2 presents resultant leakage current in the case of switching on one (phase L1), two (phases L1 and L3), or three final circuits.

The highest value of the leakage current is equal to 8 mA and occurs for operating one or two final circuits. When three circuits are operating, the resultant leakage current, due to the phase shift 120°, is the lowest: $I_{sum} = 0$ (Fig. A.2C). The resultant current should be less than the rated residual non-operating current ($0.5I_{\Delta n}$)—it allows to avoid nuisance tripping of the RCD. On the other hand, the rated residual operating current $I_{\Delta n}$ should be as low as possible (Fig. A.3). Taking the aforementioned into account, an RCD of the rated residual operating current:

$$I_{\Delta n} > 2 \cdot 8\,\text{mA} = 16\,\text{mA} \Rightarrow I_{\Delta n} = 30\,\text{mA} \text{ should be installed.}$$

Example 2

The electrical installation presented in Fig. A.4 has three 1-phase final circuits supplied from different phases (L1, L2, and L3). The following natural leakage currents flow in the final circuits:

- phase L1: resistive leakage current $I_{R1} = 3\,\text{mA}$ and capacitive leakage current $I_{C1} = 4\,\text{mA}$,

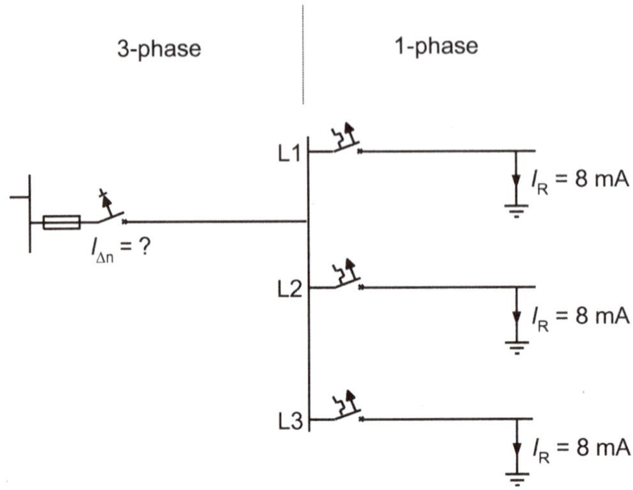

Fig. A.1 Diagram of the installation for example 1.

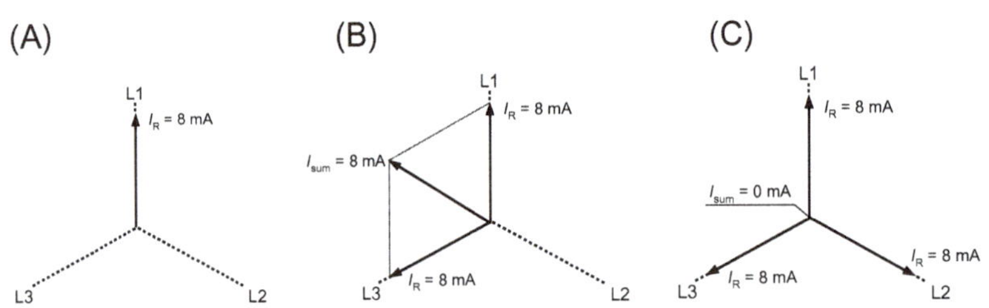

Fig. A.2 Resultant leakage current I_{sum} in the case of switching on the circuits: (A) in the phase L1, (B) in the phases L1 and L3, (C) in all phases.

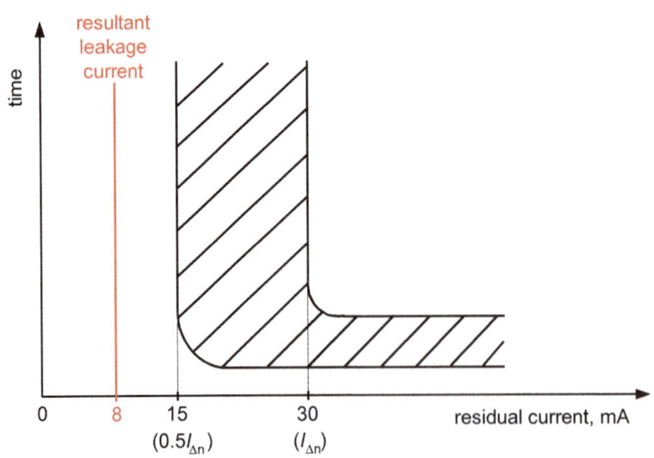

Fig. A.3 Maximal resultant leakage current in the installation from example 1 vs. time-current tripping curve of the 30 mA RCD.

- phase L2: resistive leakage current $I_{R2} = 5\,\text{mA}$,
- phase L3: resistive leakage current $I_{R3} = 6\,\text{mA}$.

Select a rated residual operating current $I_{\Delta n}$ of the RCD to be installed in the 3-phase main circuit.

Calculations and results

Leakage currents flowing in individual circuits of the analyzed installation can be presented as in Fig. A.5.

Leakage current has to be written as a complex number:

- phase L1: $I_1 = 3 + j4 = 5 \cdot e^{j\phi_1}$,
- phase L2: $I_2 = 5 \cdot e^{j(\phi_2 - 120)}$,
- phase L3: $I_3 = 6 \cdot e^{j(\phi_3 - 240)}$.

Particular circuits can be switched on in the following combinations, which gives respective modulus of a complex number:

- $|I_1 + I_2| = 0.6$ mA,
- $|I_1 + I_3| = \mathbf{9.2}$ mA,
- $|I_2 + I_3| = 5.6$ mA,
- $|I_1 + I_2 + I_3| = 5.5$ mA.

The highest value of the resultant leakage current appears in the case of the switching on circuits L1 and L3 (9.2 mA). Therefore the rated residual operating current of an RCD should be as follows (Fig. A.6):

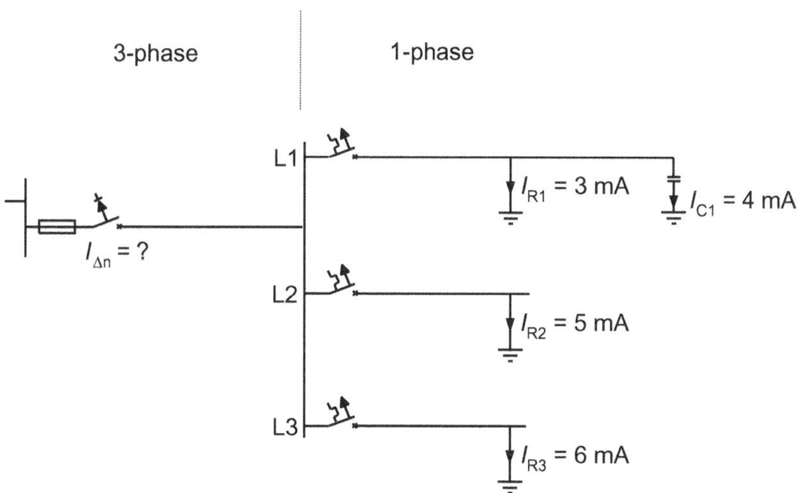

Fig. A.4 Diagram of the installation for example 2.

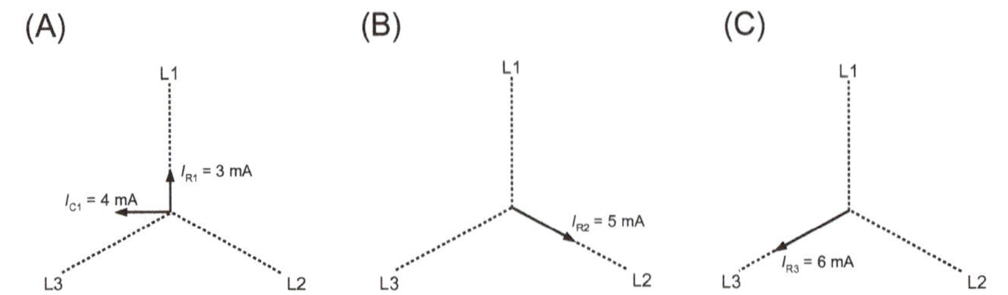

Fig. A.5 Geometrical presentation of the leakage currents flowing in the installation from Fig. A.4: (A) resistive leakage current $I_{R1} = 3\,\text{mA}$ and capacitive leakage current $I_{R1} = 4\,\text{mA}$ in phase L1, (B) resistive leakage current $I_{R2} = 5\,\text{mA}$ in phase L2, (C) resistive leakage current $I_{R3} = 6\,\text{mA}$ in phase L3.

$$I_{\Delta n} > 2 \cdot 9.2\,\text{mA} = 18.4\,\text{mA} \Rightarrow I_{\Delta n} = 30\,\text{mA}.$$

Example 3

The electrical installation has three 3-phase final circuits. Every final circuit is protected by the RCD of a rated residual operating current as specified in Fig. A.7. Select an upstream RCD to be installed in the main circuit.

Calculations and results

An upstream RCD installed in the main circuit has to be the delayed type (S-type). Its rated residual operating current should be selected with reference to the maximal possible leakage current in the main circuit generated

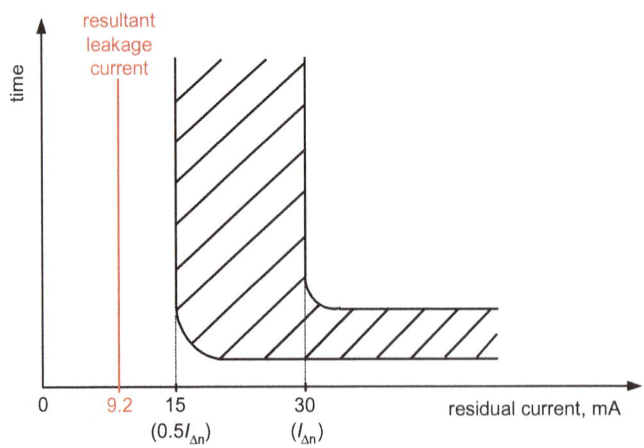

Fig. A.6 Maximal resultant leakage current in the installation from example 2 vs. time-current tripping curve of the 30 mA RCD.

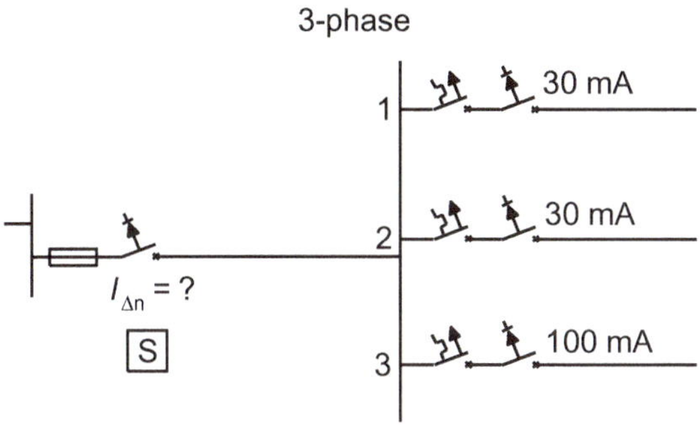

Fig. A.7 Diagram of the installation for example 3.

mainly by the final circuits. For this purpose, as a rough approximation, it can be assumed that in every final circuit the leakage current flows in the same line/phase conductor and the value of the current is equal to the rated residual operating current of the respective RCD. According to this assumption, the maximal leakage current in the final circuit is as follows:

$$I_{\Delta max} = 30\,mA + 30\,mA + 100\,mA = 160\,mA.$$

In consequence, theoretically, an RCD of $I_{\Delta n} = 500\,mA$ should be installed (Fig. A.8).

The above-mentioned assumption is very theoretical and restrictive—it may result in the selection of the RCD of too high value of the $I_{\Delta n}$. In practice, the probability of the leakage current occurrence in all circuits at the same time and of such values (almost equal to $I_{\Delta n}$ of the installed RCD)

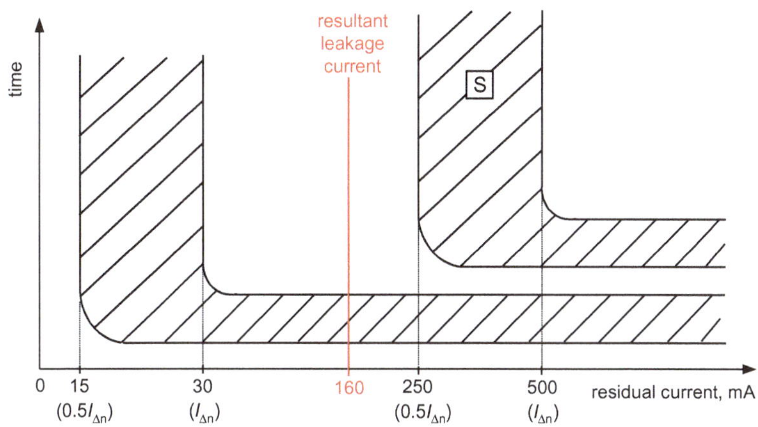

Fig. A.8 Maximal resultant leakage current (a rough approximation) in the main circuit of the installation from example 3 vs. time-current tripping curve of the 500 mA S type RCD.

is negligible. Therefore it is proposed to calculate the maximal leakage current in the main circuit according to the following procedure:

(1) Final circuits are sorted from the RCD having the highest value of the $I_{\Delta n}$ to that having the lowest value of the $I_{\Delta n}$;

(2) Such sorted values are multiplied by the respective coefficients $k_c = 1.0$, 0.8, 0.6, 0.4, 0.2, as follows:

$$I_{\Delta}\text{max} = 1.0 \cdot I_{\Delta n1} + 0.8 \cdot I_{\Delta n2} + 0.6 \cdot I_{\Delta n3} + 0.4 \cdot I_{\Delta n4} + \sum_{5}^{N} (0.2 \cdot I_{\Delta nN}) \tag{A.1}$$

where

$I_{\Delta n1} \ldots\ I_{\Delta nN}$—a rated residual operating current of the subsequent RCDs.

For the installation presented in Fig. A.7, the maximal leakage current calculated according to (A.1) is as follows:

$$I_{\Delta\text{max}} = 1.0 \cdot 100\,\text{mA} + 0.8 \cdot 30\,\text{mA} + 0.6 \cdot 30\,\text{mA} = 142\,\text{mA}$$

It allows to select an RCD of $I_{\Delta n} = 300\,\text{mA}$ (Fig. A.9).

Example 4

Electrical installation (TT system) has thirty 3-phase final circuits. Every final circuit is protected by a no-delayed AC-type RCD of a rated residual

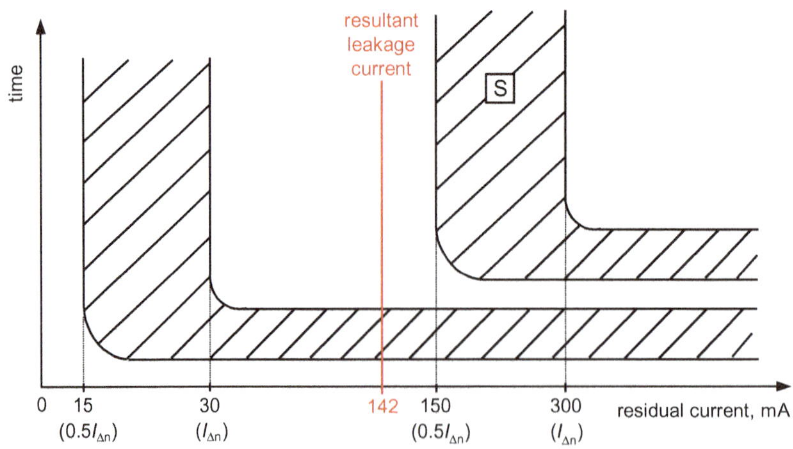

Fig. A.9 Maximal resultant leakage current in the main circuit of the installation from example 3 vs. time-current tripping curve of the 300 mA S-type RCD.

operating current equal to 30 mA (Fig. A.10). Select resistance-to-earth R_A of the main PE conductor in the installation.

Calculations and results

In TT systems, earth (residual) current flows through the earth and generates a voltage drop across the resistance R_A. This voltage drop introduces a potential to the PE conductor. For safety purposes, the voltage cannot exceed 50 V. Therefore the condition included in (4.2) has to be fulfilled:

$$R_A \cdot I_a \leq 50\,\text{V}.$$

Taking into account the tripping current of RCDs as a function of the maximum permissible disconnection time (Table 4.2), in the analyzed installation (TT system, the maximum permissible disconnection time 0.2 s,

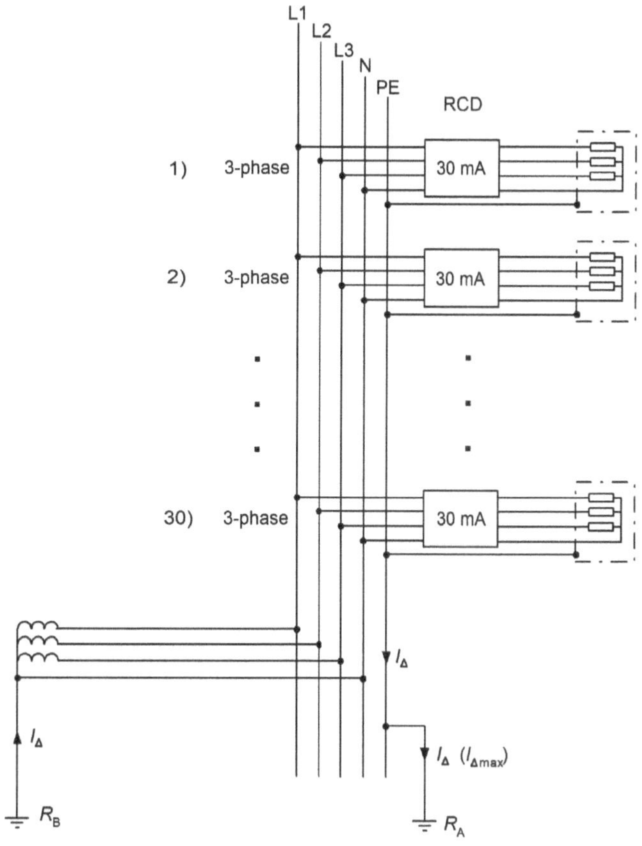

Fig. A.10 Diagram of the installation for example 4.

AC-type RCDs) the aforementioned expression can be rearranged to the following one:

$$R_A \cdot (2\, I_{\Delta max}) \leq 50\,\mathrm{V} \Rightarrow R_A \leq \frac{50\,\mathrm{V}}{2 \cdot I_{\Delta}\mathrm{max}}$$

where $I_{\Delta max}$ is the highest resultant leakage current in the installation.

The highest resultant leakage occurs when leakage currents from particular final circuits flow in the same phase (e.g., L1). Assuming that the simultaneous leakage current flow from all 30 circuits is very unlikely, expression (A.1) can be utilized to calculate the resultant leakage current $I_{\Delta max}$:

$$\begin{aligned} I_{\Delta}\mathrm{max} &= 1.0 \cdot I_{\Delta n1} + 0.8 \cdot I_{\Delta n2} + 0.6 \cdot I_{\Delta n3} + 0.4 \cdot I_{\Delta n4} + \sum_{5}^{N} (0.2 \cdot I_{\Delta nN}) \\ &= 1.0 \cdot 30\ \mathrm{mA} + 0.8 \cdot 30\ \mathrm{mA} + 0.6 \cdot 30\ \mathrm{mA} + 0.4 \cdot 30\ \mathrm{mA} \\ &\quad + \sum_{5}^{30} (0.2 \cdot 30\ \mathrm{mA}) = 240\,\mathrm{mA} \end{aligned}$$

The required resistance-to-earth R_A of the main PE conductor in the installation is then:

$$R_A \leq \frac{50\,\mathrm{V}}{2 \cdot I_{\Delta}\mathrm{max}} = \frac{50\,\mathrm{V}}{2 \cdot 0.24\,\mathrm{A}} = 104\,\Omega$$

Example 5

Electrical installation (TT system) has eight final circuits. Consecutive circuits are protected by no-delayed AC-type RCDs as specified in Fig. A.11. Select resistance-to-earth R_A of the main PE conductor in the installation.

Calculations and results

Similarly to example 4, the highest resultant leakage current $I_{\Delta max}$ in the installation has to be calculated. However, in this installation, there are both 1-phase and 3-phase circuits, and RCDs have different rated residual operating currents: 30, 100, and 300 mA. To evaluate current $I_{\Delta max}$, all circuits should be compiled as in Table A.1.

The highest value of the current has been obtained in phase L1 ($I_{\Delta max}$ = 618 mA). In the analyzed installation (TT system, the maximum permissible disconnection time 0.2 s, AC-type RCDs), the required resistance-to-earth R_A of the main PE conductor is as follows:

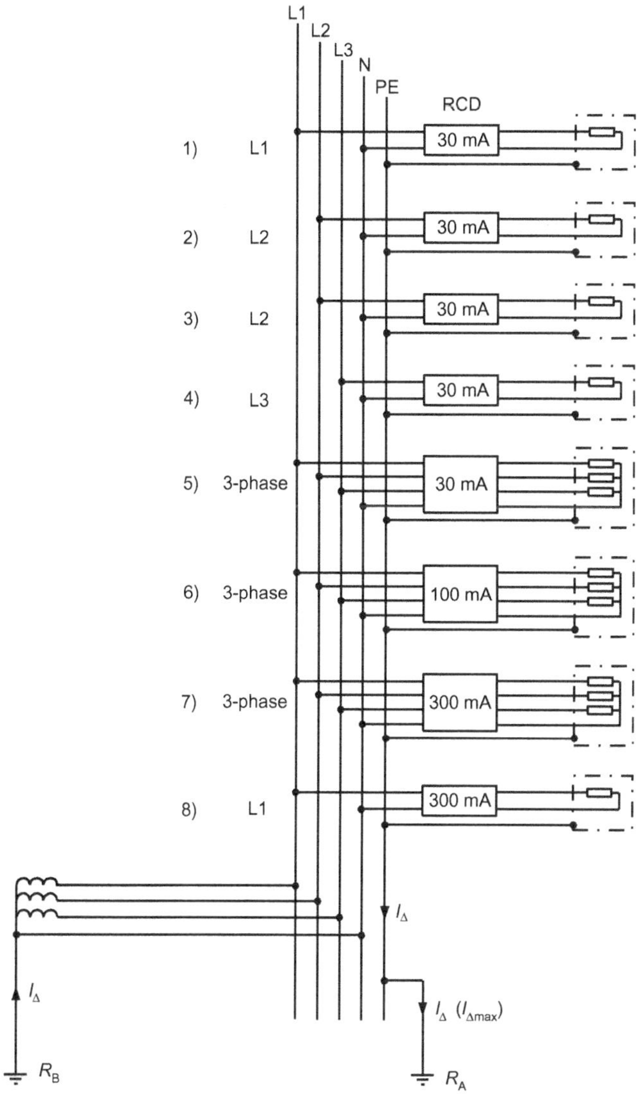

Fig. A.11 Diagram of the installation for example 5.

$$R_A \le \frac{50\,\text{V}}{2 \cdot I_{\Delta\text{max}}} = \frac{50\,\text{V}}{2 \cdot 0.618\,\text{A}} = 40\,\Omega$$

Table A.1 Compilation of the circuits according to the values of the $I_{\Delta n}$ of the RCDs.

Circuit type[a]	Rated residual operating current (mA)	Calculated leakage current (mA)		
		L1	L2	L3
3-Phase	300	$1.0 \cdot 300 = 300$	$1.0 \cdot 300 = 300$	$1.0 \cdot 300 = 300$
1-Phase (L1)	300	$0.8 \cdot 300 = 240$	–	–
3-Phase	100	$0.6 \cdot 100 = 60$	$0.8 \cdot 100 = 80$	$0.8 \cdot 100 = 80$
3-Phase	30	$0.4 \cdot 30 = 12$	$0.6 \cdot 30 = 18$	$0.6 \cdot 30 = 18$
1-Phase (L1)	30	$0.2 \cdot 30 = 6$	–	–
1-Phase (L2)	30	–	$0.4 \cdot 30 = 12$	–
1-Phase (L2)	30	–	$0.2 \cdot 30 = 6$	–
1-Phase (L3)	30	–	–	$0.4 \cdot 30 = 6$
Total		618	416	404

[a]Circuits should be sorted from the highest to the lowest $I_{\Delta n}$ (here: from 300 mA to 30 mA).

Example 6

Select the RCD type (AC, A, F, or B) to the converter circuits with the presented possible earth fault current waveform shape i_E (Fig. A.12).

Analysis and results

The circuit from Fig. A.12A contains a rectifier giving a half-wave current. Such type of current is a DC waveform with relatively high pulsation, hence, at least A-type RCD is necessary (see Table 8.2, item no. 4). Obviously, the application of F-type and B-type (B+ type) RCDs is possible as well.

The circuit from Fig. A.12B is composed of a rectifier with a smoothing capacitor. In the case of an insulation fault, the earth current i_E has relatively low pulsation. Low pulsation of the DC waveform requires to install a B-type (or B+ type) RCD (see Table 8.2, item no. 8).

The circuit from Fig. A.12C contains an L–N supplied PWM converter to control the motor. In such a circuit, high-frequency components in the earth current i_E occur. To detect composite earth/residual currents obtained in circuits supplied from single phase, application of at least F-type RCDs is necessary (see Table 8.2, item no. 7). The application of B-type (B+ type) RCDs is also possible.

The circuit from Fig. A.12D contains a PWM converter supplied from all phases. In such a variable-speed drive circuit, a DC earth current of low

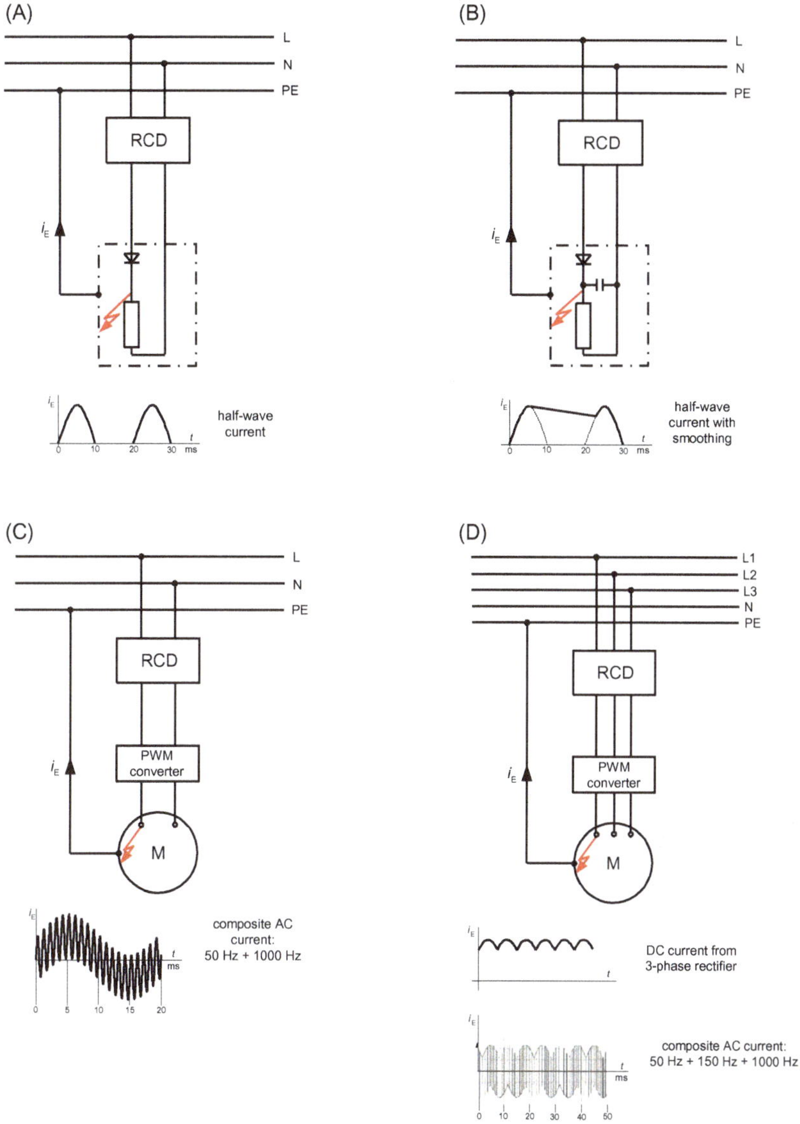

Fig. A.12 Circuit with: (A) a diode rectifier supplied L–N, (B) a diode rectifier with smoothing capacitor, supplied L–N, (C) a variable-speed drive supplied L–N, (D) a variable-speed drive supplied from all phases; i_E—possible earth fault current waveform shape.

pulsation as well as composite AC current may occur (for more details, see Chapter 8). To detect the both aforementioned types of waveform, the application of B-type (B+ type) RCDs is necessary (see Table 8.2, items no. 10, 11, and 12).

Appendix B

Standards regarding residual current devices' performance, application, and operation

1. IEC 60755:2017 General safety requirements for residual current operated protective devices.
2. IEC 61008-1:2010 Residual current operated circuit-breakers without integral overcurrent protection for household and similar uses (RCCBs)—Part 1: General rules.
3. IEC 61009-1:2010 Residual current operated circuit-breakers with integral overcurrent protection for household and similar uses (RCBOs)—Part 1: General rules.
4. IEC 62423:2009 Type F and type B residual current operated circuit-breakers with and without integral overcurrent protection for household and similar uses.
5. DIN VDE 0664-400:2012-05 Fehlerstrom-Schutzschalter Typ B ohne eingebauten Überstromschutz zur Erfassung von Wechsel- und Gleichfehlerströmen für den gehobenen vorbeugenden Brandschutz—Teil 400: RCCB Typ B+ (Residual current operated circuit-breakers type B without integral overcurrent protection to operate at residual alternating and residual direct currents for advanced preventative protection against fire—Part 400: RCCB Type B+).
6. HD 639 S1:2002 Electrical accessories—Portable residual current devices without integral overcurrent protection for household and similar use (PRCDs).
7. IEC 63024:2017 Requirements for automatic reclosing devices (ARDs) for circuit-breakers, RCBOs and RCCBs for household and similar uses.
8. IEC 62020-1:2020 Electrical accessories—Residual current monitors (RCMs)—Part 1: RCMs for household and similar uses.
9. HD 60364-4-41:2017 Low-voltage electrical installations—Part 4-41: Protection for safety—Protection against electric shock.

10. HD 60364-4-42:2011 Low-voltage electrical installations—Part 4-42: Protection for safety—Protection against thermal effects.
11. HD 60364-4-42:2011/A1:2015 Low voltage electrical installations—Part 4-42: Protection for safety—Protection against thermal effects.
12. HD 60364-5-53:2015 Low-voltage electrical installations—Part 5-53: Selection and erection of electrical equipment—Switchgear and controlgear.
13. IEC 60364-5-55:2011 Electrical installations of buildings—Part 5-55: Selection and erection of electrical equipment—Other equipment.
14. HD 60364-5-534:2016 Low-voltage electrical installations—Part 5-53: Selection and erection of electrical equipment—Isolation, switching and control—Clause 534: Devices for protection against transient overvoltages.
15. IEC 60364-5-56:2018 Low-voltage electrical installations—Part 5-56: Selection and erection of electrical equipment—Safety services.
16. IEC 60364-6:2016 Low-voltage electrical installations—Part 6: Verification.
17. HD 60364-7-701:2007 Low-voltage electrical installations—Part 7-701: Requirements for special installations or locations—Locations containing a bath or shower.
18. IEC 60364-7-702:2010 Low-voltage electrical installations—Part 7-702: Requirements for special installations or locations—Swimming pools and fountains.
19. HD 60364-7-703:2005 Electrical installations of buildings—Part 7-703: Requirements for special installations or locations—Rooms and cabins containing sauna heaters.
20. HD 60364-7-704:2007 Low-voltage electrical installations—Part 7-704: Requirements for special installations or locations—Construction and demolition site installations.
21. HD 60364-7-705:2007 Low-voltage electrical installations—Part 7-705: Requirements for special installations or locations—Agricultural and horticultural premises.
22. HD 60364-7-706:2007 Low-voltage electrical installations—Part 7-706: Requirements for special installations or locations—Conducting locations with restricted movement.
23. IEC 60364-7-708:2017 Low-voltage electrical installations—Part 7-708: Requirements for special installations or locations—Caravan parks, camping parks and similar locations.

24. HD 60364-7-709:2010/A12:2019 Low-voltage electrical installations—Part 7-709: Requirements for special installations or locations—Marinas and similar locations.
25. HD 60364-7-710:2012 Low-voltage electrical installations—Part 7-710: Requirements for special installations or locations—Medical locations.
26. IEC 60364-7-711:2018 Low-voltage electrical installations—Part 7-711: Requirements for special installations or locations—Exhibitions, shows and stands.
27. IEC 60364-7-713:2013 Electrical installations of buildings—Part 7-713: Requirements for special installations or locations—Furniture.
28. HD 60364-7-714:2012 Low-voltage electrical installations—Part 7-714: Requirements for special installations or locations—External lighting installations.
29. HD 60364-7-718:2013 Low-voltage electrical installations—Part 7-718: Requirements for special installations or locations—Communal facilities and workplaces.
30. HD 60364-7-721:2019 Low-voltage electrical installations—Part 7-721: Requirements for special installations or locations—Electrical installations in caravans and motor caravans.
31. HD 60364-7-722:2018 Low-voltage electrical installations—Part 7-722: Requirements for special installations or locations—Supplies for electric vehicles.
32. HD 60364-7-730:2015 Low-voltage electrical installations—Part 7-730: Requirements for special installations or locations—Onshore units of electrical shore connections for inland navigation vessels.
33. HD 60364-7-740:2006 Electrical installations of buildings—Part 7-740: Requirements for special installations or locations—Temporary electrical installations for structures, amusement devices and booths at fairgrounds, amusement parks and circuses.
34. IEC 60364-7-753:2014 Low-voltage electrical installations—Part 7-753: Requirements for special installations or locations—Heating cables and embedded heating systems.
35. IEC 61557-6:2007 Electrical safety in low voltage distribution systems up to 1 000 V a.c. and 1 500 V d.c.—Equipment for testing, measuring or monitoring of protective measures—Part 6: Effectiveness of residual current devices (RCD) in TT, TN and IT systems.
36. EN 50178:1997 Electronic equipment for use in power installations.
37. IEC 62640:2011 Residual current devices with or without overcurrent protection for socket-outlets for household and similar uses.

38. IEC 61851-1:2017 Electric vehicle conductive charging system—Part 1: General requirements.

39. IEC 62752:2016 In-cable control and protection device for mode 2 charging of electric road vehicles (IC-CPD).

40. IEC 62955:2018 Residual direct current detecting device (RDC-DD) to be used for mode 3 charging of electric vehicles.

41. IEC/TR 62350:2006 Guidance for the correct use of residual current operated protective devices (RCDs) for household and similar use.

42. IEC 62109-1:2010 Safety of power converters for use in photovoltaic power systems—Part 1: General requirements.

43. IEC 63024:2017 Requirements for automatic reclosing devices (ARDs) for circuit-breakers, RCBOs and RCCBs for household and similar uses.

Index

Note: Page numbers followed by *f* indicate figures, and *t* indicate tables.

FSC
www.fsc.org
MIX
Papier | Fördert gute Waldnutzung
FSC® C083411

Kolibri 360 GmbH
Ferdinand-Jühlke-Straße 7
99095 Erfurt, Deutschland
produktsicherheit@kolibri360.de